Making Sense of Science

Making Sense of Science

Separating Substance from Spin

Cornelia Dean

The Belknap Press of
Harvard University Press

Cambridge, Massachusetts
London, England
2017

SECOND printing

Library of Congress Cataloging-in-Publication Data

Names: Dean, Cornelia, author.
Title: Making sense of science : separating substance from spin / Cornelia Dean.
Description: Cambridge, Massachusetts : The Belknap Press of
    Harvard University Press, 2017. | Includes bibliographical references and index.
Identifiers: LCCN 2016037738 | ISBN 9780674059696 (alk. paper)
Subjects: LCSH: Science news. | Research—Moral and ethical aspects. |
    Research—Political aspects. | Science in popular culture.
Classification: LCC Q225 .D43 2017 | DDC 500—dc23
    LC record available at https://lccn.loc.gov/2016037738

In memory of my mother,
WINIFRED O'MALLEY DEAN

"Nothing is more important to a democracy than a well-informed electorate."

—Will McAvoy, managing editor, Newsnight, Atlantis Cable News (*Newsroom,* episode 3, by Aaron Sorkin, HBO)

CONTENTS

For thirty years I have been a science journalist at the *New York Times*; for almost seven years, I headed the newspaper's Science Department, supervising coverage of science, medicine, health, environment, and technical issues. From Day One, I have thought that covering these subjects for the *Times* was the best job in the world.

As science editor I dispatched correspondents to cover research in every continent. And I traveled myself, to places like the Dry Valleys of Antarctica, where researchers study what may be Earth's simplest food web, and Palomar Mountain, in California, where the 200-inch Hale Telescope gathers up the light of the universe.

We had the intellectual support of the editorial hierarchy in a newsroom that in those days was flush with money. For example, I once took it upon myself to dispatch a reporter to the Chilean Andes when astronomers there observed a newborn supernova. Coming from a newspaper where long-distance telephone calls were subjects of budgetary debate, I found this kind of thing thrilling.

Today, the money situation is not so good. The business model of the mainstream "legacy" news media is broken, and we do not know what will fix it. Though there are bright spots—like the *Times,* where support for science coverage is still strong—news outlets generally are cutting back on science and environmental reports. Across the country reporters on science and environment beats have been reassigned or fired outright. Medical coverage too often focuses on "news you can use" clickbait rather than advances in understanding human biology.

Internet sites take up some of the slack. But while the web can be a valuable source of information, it propagates bad information and even deliberately fake news along with the good. And the toxic quality of many comment threads online can actually turn people away from high-quality science news. At least, that's what researchers from the University of Wisconsin found in 2013. The finding was one of the reasons the magazine *Popular Science* ended the practice of allowing readers to comment online on its articles, as its online content director noted on September 24, 2013.

So while I continue to write for the *Times,* I have been spending more and more of my time talking to scientists and engineers about the need for them to communicate with the public. I teach seminars for graduate students and postdoctoral researchers to help them develop the skills they need to conduct that kind of engagement successfully. I participate in training programs for young researchers seeking to engage more fully in the debates of the day. This engagement, I tell them, is (or should be) part of what it means to be a researcher.

In 2009 I distilled some of this thinking into a book, *Am I Making Myself Clear?,* aimed at encouraging researchers in these efforts. The book you are reading now is aimed at the other important part of the equation: the public, citizens, *us*—the people who struggle to make sense of what we hear when scientists and engineers talk to us and—especially—when others try to shape or spin those messages to their own advantage.

Of course, this book is written by a journalist, not a researcher. Even worse, it is the book of one journalist—me—and I have no scientific or other technical training of any kind. I was first transferred into the *Times*'s science department, on what was supposed to have been a temporary assignment, because someone (true story) had seen me in the newsroom carrying a copy of *Scientific American.*

So my goals are relatively modest. I hope to show my readers the kinds of thinking we do in the newsroom when we try to decide whether a given finding is newsworthy, trustworthy, and important. I will illustrate the book with examples from my experiences as a science journalist. Inevitably, I will leave things out or, I fear, ascribe undue importance to things just because I experienced them. Still, I hope the material I have collected here will help people assess the scientific and technical claims and counterclaims they increasingly encounter in public debates and in their private lives.

Chapter 1 will describe the problems we ordinary folk bring to this table—our ignorance, irrational patterns of thinking, inability to think probabilistically, and erroneous ideas about risk. Chapter 2 will discuss the research enterprise—what science is (and isn't), how the scientific method works today, the use of computer models in research, and the sometimes problem-plagued process of peer review. Chapter 3 will talk about what happens when things in this world go wrong, either because researchers have misbehaved or because they find themselves in unfamiliar arenas, in courtrooms or quoted in the media.

The second part of the book will describe how all these factors play out. Chapter 4 will discuss how the research enterprise is financed, and how money influences things like our health care and what we eat. Chapter 5 will discuss the influence of politics on the conduct of science, particularly when it comes to the environment and arguments over religion.

The Appendix offers nuts-and-bolts advice on assessing who is an "expert," reading research papers, and deciding whether to believe polls and surveys.

This book cannot possibly be a comprehensive guide. It will not provide all the answers, or even many answers. It will, however, provide some interesting and (I hope) useful questions.

Making Sense of Science

INTRODUCTION

We live in an age of science. Daily, we learn more and more about the biological and physical world. But ignorance and untruth are hallmarks of our times as well. People with economic, electoral, or ideological agendas capitalize on our intellectual or emotional weaknesses to grind their own axes. Too often, they get away with it.

Examples of distortion or outright deception are numerous and come from all sides of the political spectrum. Industries persuade government agencies to write regulations that suit their needs rather than the needs of the public or the environment. Industry lobbyists may even write them. And politicians hungry for campaign contributions go along with it. Advocacy organizations, hoping to attract new members and contributions, may fight these industries with loud but exaggerated or even invalid warnings about supposed dangers to public health or the environment.

Science can be misused by doctors, who may prescribe drugs or order tests without proven benefit; by religious leaders, who spread misinformation to support their doctrines; and by politicians and lobbying groups, among others, who court voters aligned to their goals.

How can all this happen? The answer is simple. Politicians, lobbyists, business interests, and activists make their cases in the public arena, back their arguments with science or engineering "facts," and rely on the rest of us to leave them unchallenged. Most of us don't have the knowledge or the time to assess scientific and technical claims. It may not even occur to us that assessing such claims is something we ought to try to do.

The issues we face today are important and complex, but they will pale before the technical, ethical, and even moral questions we will confront in the not-too-distant future. Should scientists inject chemicals into the atmosphere or the ocean to counter global warming by "tuning" Earth's climate? Should people be able to design their own children? Should engineers design battlefield robots that decide, on their own, when to fire their weapons, and at whom? Work on these and other similarly game-changing technologies is already underway. Some of these technologies may be beneficial, even vital. But all should be debated and discussed by a knowledgeable public.

Are we up to it? It's hard to say confidently that we are. Science in the United States is taught poorly; knowledge of statistics is hardly taught at all. As a result, many of us are ignorant not just of scientific and engineering facts but also of the ways the research enterprise gathers its facts. Without the necessary skills to assess the data, we struggle with statistics and embrace downright irrational ideas about risk.

Nevertheless, we admire the research enterprise—a lot. In survey after survey, scientists, engineers, and physicians rank among those earning the highest respect from Americans. That is why people wishing to advance their own ends often seek to clothe their arguments in the garments of research, asserting "the science is on my side." We may accept these arguments simply because they sound credible to our untutored ears and, more importantly, because they mesh with our political or economic views or resonate with some fear or prejudice we may not even realize we hold.

Of course, a certain amount of irrationality is nothing new in the nation's political and economic life. But when the question is raising local property taxes or cutting Medicare benefits, most people have some intuition or life experience to guide them. While there is much that is arcane in economics, for example, most people know, more or less, whether their job is secure, if they have enough money to get by, or if people in their town are doing better or worse than they were a year ago.

That is not the case with science. Often its findings are counterintuitive or downright bizarre—as when researchers tell us hairspray aerosols erode the protective ozone layer over the South Pole, or that reducing acid rain by clearing the air of sulfur dioxide pollution can make the planet warmer. The intuitions we have about science often point us in wrong directions.

Meanwhile, as the federal government starts to step back as a backer of scientific research, the profit motive increasingly determines what is studied, how studies are designed—and whose findings become widely known and whose results are buried. All of these trends seem likely to worsen in a Trump presidency.

Two groups of people could help us separate fact from hype: researchers and the journalists who report on their work. But the culture of science still inveighs against researchers' participation in public debates. With rare exceptions, scientists and engineers are absent from the nation's legislatures, city councils, or other elective offices. Their training tells them to stay out of the public eye even when they have much to say that could inform public debates. In effect, they turn the microphones over to those who are unqualified to speak. Though I keep hearing that this institutional reserve is cracking, I do not believe it has cracked enough. And mainstream journalism, the kind of reporting that aims to give people the best possible approximation of the truth, "without fear or favor" (a credo of journalists at the *New York Times*), is struggling. Until we figure out how to fix it, journalists will often lack the financial resources they need to do a good job with complex technical subjects.

The result is a world in which researchers gather data; politicians, business executives, or activists spin it; journalists misinterpret or hype it, and the rest of us don't get it. Whoever has the most money, the juiciest allegation, or the most outrageous claim speaks with the loudest voice. The internet, newspapers, the airwaves, the public discourse generally are all too often brimming with junk science, corrupt science, pseudoscience, and nonscience.

More than fifty years ago, the British chemist and novelist C. P. Snow gave a speech at Cambridge University, in England. His subject was what seemed to him to be the greatest intellectual challenge of his age, a vast gulf of mutual incomprehension widening between scientific and literary elites, between science and the humanities. "In our society (that is, advanced Western society), we have lost even the pretence of a common culture," he said. "Persons educated with the greatest intensity we know can no longer communicate with each other on the plane of their major intellectual concern. This is serious for our creative, intellectual and, above all, our normal life. It is leading us to interpret the past wrongly, to misjudge the present, and to deny our hopes of the future. It is making it difficult or impossible for us to take good action."[1] His speech electrified his audience and set off a long, loud public debate on both sides of the Atlantic.

As he saw it, scientists did not know enough or care enough about the arts and literature of their own societies. And the literary elites knew so little about science and technology that they could not recognize genuinely stunning changes in our understanding of the natural world. Each group scorned the other as a bunch of ignorant specialists. In a kind of Luddism that persists today, literary intellectuals dismissed many of the benefits of science and technology, and longed for a more primitive and supposedly "genuine" age when people lived natural lives, closer to the land. For Snow, this kind of reasoning was idiotic. "It is all very well for us, sitting pretty, to think material standards of living don't matter all that much," he wrote.[2] But they do, and he had the parish records, ancient census reports, and other data to prove it: when people could choose to remain in technologically backward rural Edens or take jobs in factories, the dark satanic mills won out every time.

Much has changed in the decades since Snow's speech. Today, the gulf is not so much between scientific and literary elites as between scientific and engineering elites and everyone else. Few speeches by any scientist (or novelist) at any university electrify anyone nowadays. In itself, that can be read as evidence of the way academics have abandoned the public arena or even the ordinary world of everyday life.

True, science is in our culture today to a much greater extent than it was in 1959. To cite just two examples, computer science and medical research have transformed our lives, and everyone knows it. But though we may know how to surf the internet (even if we don't know how the information travels to our screen), and we may demand that every twinge be explored with an MRI (even if we don't know how the imaging machine works), the conduct of research, the scientific method, the *process* of discovery—is as mysterious to most of us as it was when Snow spoke.

In addition, as a society we seem much more willing than we once were to simply ignore or dismiss inconvenient scientific facts. Whether the issue is energy policy, stem cell research, cancer testing, or missile defense, scientific and engineering progress seems paradoxically to have left us more, not less, vulnerable to spin. If anything, the nonscientists of today are even more ignorant of science than were the literary intellectuals of five decades ago. Perhaps Snow could see this trend coming. In his original speech, when he wanted to illustrate their great ignorance, he said many of them could not even describe the second law of thermodynamics. Perhaps because too many of his readers proved his point, he changed the reference in a later version, saying merely that too many people are unfamiliar with the ins and outs of microbiology.

Our ignorance of basic scientific principles is troublesome, but much worse is our frequent inability to distinguish between science, pseudoscience, and outright hoaxes. Robert Park, then a physics professor at the University of Maryland, experienced the phenomenon when he served as the representative of the American Physical Society in Washington, D.C.—in effect, the lobbyist for the nation's physicists. "Of the major problems confronting society—problems involving the environment, national security, health, and the economy—there are few that can be sensibly addressed without input from science," he wrote a few years ago.[3] "As I sought to make the case for science, however, I kept bumping up against scientific ideas and claims that are totally, indisputably, extravagantly wrong, but which nevertheless attract a large following of passionate, and sometimes powerful, proponents. I came to realize that many people choose scientific beliefs the same way they choose to be Methodists, or Democrats, or Chicago Cubs fans. They judge science by how well it agrees with the way they want the world to be." This phenomenon is called "cultural cognition" or "motivated reasoning," and it is a subject of lively research and debate. It has a lot to say about how we respond to issues as diverse as climate change and gun control.

Meanwhile, many people positively flaunt their ignorance of science. When I was the science editor at the *New York Times,* it was not unusual to hear even senior colleagues proclaim proudly that they could not fathom mathematics, biology, physics, or any number of other technical subjects. That is not the kind of admission any of them would ordinarily make about politics, economics, bicycle racing, wine, military strategy, or poetry, regardless of the depths of their ignorance. Pride in technical ignorance is not something we can afford as a society.

Distrust of all major public institutions is on the rise. Since the Reagan administration, the phrase "public service" has become a contradiction in terms. Congress, state legislatures, and other governing bodies are scandal-ridden and so riven by politics that they are hardly able to function. In the wake of clergy sex scandals, even religious groups are losing influence.

We might, in theory, fall back on ourselves, on the so-called "wisdom of crowds." But though crowds may do better than individuals in estimating the number of jellybeans in a jar or the weight of an ox, crowd-sourcing on an issue less specific works best (or maybe only) when there is wide agreement in the crowd. When people disagree, all too often they attempt to shut opposing views out. That's why Wikipedia is most reliable on matters on which there is little to argue about—and why, if there are arguments, the site sometimes has to shut down a topic page.

Where does all this leave us? On our own. We can cross our fingers, hope for the best and declare, like climate-dodging politicians, "I am not a scientist!" Or we can try to understand the issues that confront us, as deeply—and as usefully—as we can.

1

WE THE PEOPLE

What We Know, and What We Don't Know

My *New York Times* colleague Claudia Dreifus regularly interviews researchers; edited versions of her "Conversations" have been running in the weekly Science Times section since 1998. I recruited Claudia for this job because I admired her interviews with other kinds of people—political figures, for example—and because I was eager to let readers of the science section hear scientists speak in their own voices, not just in a sound bite here and there but in an extended back-and-forth.

Claudia's interviews were a great success, and in 2002 the *Times* collected them in a book.[1] Soon after, Claudia gave a reading to promote the book, and I was in the audience. She described her initial worries about whether, given her lack of science training, she would be able to do a good job interviewing scientists. She learned, she said, that "science is very interesting if you get over the idea that you can't get it."[2]

She's right, but too many people don't realize it. Many of us don't know much about science, and we assume—probably because of bad experiences in high school or college science classes—that we "can't get it," that we will never know much about science or engineering, that the subjects are just too complex for us.

As a result, we don't engage. As a result of that, we are ignorant of the facts, we don't know how to think about what we *do* know, and we develop irrational ideas about all kinds of things, including risk. So our first task in assessing scientific or technical claims is to take inventory of our own mental defects and consider how they get in our way and how we can work around them.

One of our problems is ignorance. As a group, we Americans don't know a great deal about science. Jon Miller, a researcher at the University of Michigan who for decades has been surveying people in the United States and abroad on their knowledge of and attitudes toward science, once told me he estimates that only about a quarter of Americans are, as he put it, "scientifically literate"—able to read and understand the Science Times section of the *New York Times*. I don't think he is setting the bar very high. (His assessment also disappointed me because the writers and editors of that section strive to produce copy that almost any curious nonexpert reader can understand.)

Be that as it may, in many surveys we Americans don't show up too well. For example, periodic surveys assessed by the National Science Foundation show many of us don't know that atoms are smaller than molecules or even that Earth moves around the Sun and takes a year to do it. Only a minority of Americans accept the theory of evolution, the most abundantly supported idea in all of science.[3]

On the other hand, we are remarkably ready to accept utter nonsense. The survey finds widespread belief in the occult—about a third of us believe astrology is at least "somewhat" scientific. That's probably because we are far more likely to remember predictions that come true than the ones that don't. Perhaps it's also because far more newspapers run daily astrology columns than run any science columns at all. And it explains why, in most bookstores, titles on flying saucers far outnumber books about flying machines.

For years, people in and out of the STEM (science, technology, engineering, and math) community have bemoaned what they call the poor quality of science education in the United States. In 1983, in "an open letter to the American people," the National Commission on Excellence in Education spoke with alarm about the deficiencies in the nation's elementary and high school educational system.[4] In 1985, the American Association for the Advancement of Science began Project 2061, an effort to improve science education, including periodic assessments of what students actually know.[5] And in 2010 the President's Council of Advisors on Science and Technology (PCAST) issued a report: "Prepare and Inspire: K–12 Education in Science, Technology, Engineering and Math (STEM)."

As others had in the past, PCAST cited the importance of scientific and technical proficiency for people entering the job market, and for the nation's economic growth. The report called for five steps—training additional accomplished STEM teachers; rewarding effective teachers with extra "merit" pay and other benefits (an idea long unpopular among unionized teachers); establishing a new federal agency to develop instructional materials (an idea sure to encounter opposition from creationists and others on the religious right); supporting state efforts to establish STEM-focused schools; and taking steps to establish after-school STEM-related programs.

Has anyone heard these messages? Not enough of us, apparently. In January 2011, the Department of Education reported that only a fifth of the nation's high school seniors were what the department called "proficient" in science, the second-lowest level of any subject area covered by the National Assessment of Educational Progress, a test administered to a sample of 11,000 twelfth graders in 2009.[6] (Their performance in history was even worse.) Only one or two percent demonstrated enough mastery of science to be called "advanced."

What is the problem? For one thing, people with math, engineering, and science degrees are, even in a down economy, highly employable, usually at above-average salaries. Though there are exceptions, of course, persuading these people to take jobs in the nation's math and science classrooms can be a hard sell. Many technically proficient people also don't want to jump through the requisite teaching certification hoops.

In far too many high schools, science and mathematics are taught not as fascinating fields filled with important problems to be solved, but rather as collections of facts to be memorized and regurgitated. Science labs are not places where the quest for knowledge is carried out but rather places where students are given a set of materials and instructions to conduct "experiments" whose results are known in advance. They are told ahead of time what their lab work will produce. This cookbook approach is the antithesis of science. And it is not much fun.

According to an expert panel convened by the National Research Council to assess the situation, American high school students don't ordinarily ask their own questions, figure out their own experiments to answer them, or consider what their results have to say about natural phenomena. Plus, the report said, teachers are rarely trained to teach labs, class schedules are not conducive to lab work, and the pressure to adhere to testing guidelines means labs often get short shrift.[7] "Even the most artfully designed inquiry-based lab . . . must compete for time in a crowded academic schedule," the journal *Science* wrote in assessing the panel's report.[8] So many students emerge from high school or even college viewing science classes as alternately humiliating or boring.

Another problem is the typical high school science curriculum: biology followed by chemistry and then physics. This pattern of instruction came into wide use at the beginning of the twentieth century, just as more and more young people started continuing their education beyond the eighth grade. At that time, biology was regarded as the easiest science to learn. It was thought of as a largely "descriptive" activity. So it came first. Physics, regarded as the most complex, came last. Today, many experts agree this approach is backwards.

Now, there is growing recognition that biology is the most complex scientific field. Understanding biology depends on having an understanding of chemistry, which in turn depends on having an understanding of physics. In an ideal world, advocates of change say, high school students would routinely learn physics first, then chemistry, then biology. (Optimists in this crowd envision a day when students would learn physics in ninth grade, then chemistry, then biology, and then, as seniors, would take a more complex physics course.)

But there is one immediate problem: teaching ninth or tenth graders physics means teaching younger students enough math to make the experience worthwhile. Inspired middle school math teachers are not exactly thick on the ground.

In 2002, I thought beliefs and attitudes might be changing, at least as far as evolution was concerned. That year, the NSF reported, for the first time, that a majority of Americans—a bare majority, 53 percent—accepted the theory of evolution. The agency announced this finding with a celebratory press release. At last, evolution had won over a majority of Americans!

But when I asked Miller about the survey results, he dismissed them. He said they were an artifact, the result of news accounts of an uproar then underway in Kansas over the state school board's decision to add creationism to the state's high school biology curriculum.

Sure enough, within a couple of years, the survey responses dropped back down to a mere 45 percent accepting evolution. Eventually, the NSF stopped considering evolution in asking questions about scientific literacy, calling the issue too charged.

The teaching of evolution continues to be contentious. Usually, we hear about it when someone sues school authorities somewhere over the issue. Meanwhile, though, in school districts all over the country, evolution is quietly dropped or glossed over in biology instruction, perhaps because, according to the National Science Teachers' Association, about a third of the nation's science teachers are creationists themselves.[9]

But evolution is the foundation upon which the modern edifice of biology and medicine is built. If students need to emerge from high school with a decent understanding of biology, they must understand evolution—and they must understand why creationism is a religious idea, not a competing scientific theory.

Education about the environment is another area in which curriculum standards are under attack. Private industry groups advocate for the addition or omission of information within state curricula. For example, the coal industry produces its own curricular materials emphasizing coal's usefulness and downplaying its many negative environmental effects.

A 2008 survey of college students about climate change found widespread ignorance, whether or not they believed the change is real. They attributed it to a hole in the ozone layer (the phenomena are unrelated), were disinclined to believe human activity is the main culprit (it is) and worried that melting sea ice would cause coastal flooding (melting of floating ice will not raise sea levels, though melting of glaciers and inland ice sheets will).

"Student misconceptions about . . . climate change have been documented at all educational levels," a geologist at the University of St. Thomas in St. Paul, Minnesota wrote in *EOS,* the transactions of the American Geophysical Union.[10] In large part, he attributed their wrong ideas to "climate myths and misinformation that are perpetuated by a small but vocal group of politicians and climate change skeptics."

Meanwhile, another survey points to the nation's science teachers as possible sources of misinformation on climate. "Notably, 30 percent of teachers emphasize that recent global warming 'is likely due to natural causes' and 12% do not emphasize human activity" as a cause of climate change.[11] Sometimes, the researchers said, the teachers adopt this stance because of political pressure in their communities. But sometimes they don't know very much about the subject, particularly the overwhelming degree to which the world's climate scientists agree that human activity, chiefly the burning of fossil fuels, is behind the problem. Of course, science and engineering issues are not the only centers of ignorance among American youth. Surveys show that many cannot name the three branches of government, don't know why Abraham Lincoln was an important figure, and cannot find the Pacific Ocean on a map.

Although, as a group, American students do not perform well when they are compared with students from other countries in tests or competitions in math or science, results on these tests depend a lot on factors, such as the size of a nation's student body, or what percentage of its students are eligible to enroll in science or engineering courses, even in high school.

Also it is hard to ignore the fact that when the world faces a scientific or technical problem—everything from the Ebola virus to tsunami detection—eyes turn to the United States. We lead the world in research science and engineering. Unfortunately, scientific and technical elites performing superbly will only take you so far in a democracy. And improving science education alone is not enough to meet our democracy's need for an informed citizenry. Many of the issues we confront now as voters are entirely new to us. Stem cell research, carbon cap and trade, artificial life—were these topics covered in your high school science classes? Probably not, even if you are under thirty. But they are issues you will confront as a voter. So the issue is not increasing the number of scientific or engineering "facts" stored in students' heads by the time they leave high school. Rather, it is a matter of teaching them how to assess new claims and findings.

And while widespread ignorance is bad, what is worse in a democracy is the positive embrace of ignorance we see in some political circles. Far from seeking knowledge about pressing questions like climate change, the safety of the food supply, or the utility of gun control laws, some of us form fixed opinions *before* doing the research, and are unwilling to seek information that might contradict those opinions. Still worse, some even characterize scientists who study these questions as out of touch with ordinary people.

So improving science education in the schools may be necessary, but it will never be sufficient. There are always going to be things we will need to learn as adults. We need practice how to learn, and we need to acquire patterns of thinking that allow us to consider all sides of an argument dispassionately. Unfortunately, those are traits many of us lack.

The Belief Engine

A few years ago, researchers reported that hurricanes with female names were deadlier than storms with male names.[12] Is that possible? No, even though two of the most destructive storms in recent years were named Katrina and Sandy. In surveys, the researchers found, the participants were more likely to say they would evacuate a coastal area in advance of a storm with a male name than one with a female name. The female names seemed less frightening. Between 1950 and 2012, the researchers noted, storms with female names caused on average forty-five deaths; storms with male names caused on average twenty-three deaths. (The National Hurricane Center did not start giving storms male names, alternating with female, until 1979, by which time the coast was much more heavily developed, and therefore more vulnerable.)

This study points to something important: people don't necessarily judge scientific information—like meteorological reports on approaching storms—according to the facts. While ignorance of scientific or engineering reality is a central problem, it is only one of the factors that lead us to bad decisions—like staying put in a dangerous place when a storm approaches. Many other factors come into play: shortcuts in thinking we use to navigate the world, often with undesirable results; our unwillingness to accept information not in accord with opinions we already hold; wild misunderstandings about statistics and downright irrational ideas about risk. These patterns of thought are so persistent and so widespread that they seem to be hard-wired in us, to the point that the ordinary human brain is not a sharp analytical instrument but rather a credulous receptacle for erroneous ideas—what the physicist Robert Park calls "a belief engine."[13]

The idea that people are, on the whole, so out-to-lunch they cannot be good citizens is not new. It was a theme of the political commentator and theorist Walter Lippmann who made the point in his 1922 essay "Public Opinion." He said the habit of clinging tenaciously to irrational ideas hindered people's ability to make rational decisions. Did Lippmann really mean to issue so sweeping an indictment? It's hard to say. But there is no doubt that the human mind is a sink of irrationality.

Over the years, many researchers have attempted to plumb its depths. Two of the most successful were the psychologists Daniel Kahneman and Amos Tversky, who collaborated for years to discern ways our human brains go off the rails when we have to make judgments, especially in uncertain times. Tversky and Kahneman called these thought patterns heuristics (roughly, rules of thumb) and biases. They described them in 1974 in a paper now regarded as a landmark of science.[14]

Kahneman, who won the Nobel Prize for this work in 2002, elaborated on the work in his 2011 book *Thinking, Fast and Slow*.[15] Kahneman, Tversky, and other researchers have identified a long list of irrational patterns of thought that plague us as a species. Here are some examples. Some are almost amusing. Others have serious consequences. Think about the degree to which you fall into these patterns. For me, the answer is: often.

PRIMING

If you ask people to name an animal, roughly 1 percent will say "zebra." If you ask people first where Kenya is and then what are the opposing colors in a game of chess, and *then* to name an animal, roughly 20 percent (including me) will say "zebra."[16] We are responding to what experts call *priming*.

ANCHORING

Researchers have also found that if you ask people to write down the first three digits of their telephone number, and then ask them to guess when Genghis Khan died, they will be more likely to say he died before the year 1000—in a three-digit year—than if you don't make the phone request before you ask the question. (He died in the 1200s, possibly in 1227.)

GENERALIZING FROM EXAMPLES

We accept large conclusions drawn from small amounts of data. Journalists, in particular, must be wary of this habit—it is surprising how often researchers will make sweeping claims about the action of a drug, the effects of an environmental pollutant, or some other phenomenon on the basis of only a few good data points. Kahneman and Tversky called this "insensitivity to sample size." Often people get away with these generalizations. Few people look deeply (or even

16

superficially) into the data and anyway, for most of us, a good story means more than a barrel of evidence. If someone offers us a good story, we don't necessarily even look for the evidence.

ILLUSORY CORRELATION
The "belief engine" is always seeking to derive meaning from the information it absorbs. It regularly sees patterns or connections in what are really random occurrences or bits of data, a phenomenon known in psychology as *apophenia*. Sometimes, for example, the brain assumes that if A happens and then B happens, A must have caused B. Once the brain has constructed a belief like that, it begins looking for support for it, often blinding itself to evidence that contradicts it.

Lawyers (and logicians) call this thinking *post hoc ergo propter hoc,* from the Latin for "after that, therefore because of that." Stated in such bald terms, the reasoning is obviously faulty, but it is stunning how often people infer causality from patterns of correlation.[17]

One of the most flamboyant examples of this thinking was the litigation that erupted over silicone breast implants. By the early 1990s, thousands of American women had received these breast implants, some for reconstruction after cancer surgery but the majority for cosmetic reasons. Doctors had known for some time that the implants could produce localized problems, like scarring, and that they could even rupture inside the body. Now, though, they were hearing far more alarming reports. Women with silicone implants were suffering systemic diseases like multiple sclerosis or lupus that were leaving them permanently disabled. Soon, there were lawsuits all over the country. Merrill-Dow, the leading silicone implant maker, was under siege. Eventually, a class action lawsuit resulted in the creation of a $3.4 billion trust fund to compensate the women who had been disabled by their implants.

There was only one problem. There was never any evidence that the implants or the silicone they contained caused the women's disabilities. There were plenty of doctors prepared to testify about how sick their patients were—they *were* sick. Scientists hired by the plaintiffs and their lawyers produced studies linking their ailments to their implants. But the women with implants had diseases thousands of American women contract every year, and at rates no higher than average. Implant manufacturers, who had repeatedly made this point in litigation, were finally vindicated when a scientific panel, appointed by the judge hearing the class action suit, declared emphatically that there was no reason to believe the implants were at fault.

By then Merrill Dow was bankrupt, and the $3.4 billion had flowed from implant makers to plaintiffs and their lawyers. New regulations had barred the use of silicone not only in breast implants but also in a host of other useful medical devices.

Education and high income do not necessarily protect us against this kind of irrationality. Compared to the national average, people in the San Francisco Bay Area are wealthier and better educated, yet the Bay Area is a center of opposition to childhood vaccination, motivated by the erroneous belief that vaccines have been linked to autism.

Initially, the idea was that mercury, a vaccine preservative, must somehow kill nerve cells in susceptible children. There has never been any evidence for this idea, but because autistic children typically receive the diagnosis in early childhood, after rounds of routine vaccinations, it has taken hold. It persists despite the fact that autism rates continued to climb after mercury was removed from vaccines, and also despite the accumulating evidence that children with autism suffer not because of brain cell death but because their brains do *not* undergo a kind of cellular pruning process that normally occurs at about age two. Parents who refuse to vaccinate their children in a timely manner have created conditions that fuel outbreaks of illnesses like whooping cough and measles.[18]

Post hoc ergo propter hoc may be a notoriously erroneous concept, but it is hard to fight. A story that taps into our emotions is much more likely to be remembered than a demographic study. But some irrational mental glitches are far less obvious.

AVAILABILITY

We put a lot of emphasis on things we have seen and experienced or observed ourselves. For example, we may choose a particular medical treatment because a relative had it, or even buy a stock because someone we know has made money with it.

In their landmark paper, Kahneman and Tversky speculated that this pattern of thought may be an evolutionary holdover from the days when there was hardly any such thing as "data." What happened to people who lived in your immediate surroundings was almost certainly more relevant to you than what happened to people elsewhere.

FRAMING

Suppose a deadly epidemic has broken out and the disease is expected to kill 600 people. Which drug is better: Drug A, which will save 200 people for sure, but only 200 people; or Drug B, which has a 1/3 probability of curing everyone and a 2/3 probability of saving no one? Given this choice, most people will choose Drug A, the drug that will certainly save 200 people. Yet if Drug A is described as dooming 400 people for sure, most people choose Drug B. Other surveys have shown that if doctors present a surgical procedure as having a 10 percent mortality rate, most people will reject it. If they say it has a 90 percent survival rate, most people will accept it. These two scenarios illustrate something psychologists call *framing effects*; that is, the degree to which the way a question is framed determines how we will answer it.

Here's another example: obesity in the United States. Is it a problem of individual appetites out of control? Maybe it is a result of misguided agricultural subsidy policies that have produced vast corn surpluses that in turn produce vast supplies of high-fructose corn syrup, a staple of inexpensive (and fattening) processed food products. Or maybe, as some activist groups tell us, obesity is nothing like the health menace we have been told it is, and the constant drumbeat about it is little more than bigotry.

Framing effects have always been important in policy making but today they are more important than ever, especially with science- or engineering-related issues whose details may be unfamiliar, complex, or otherwise hard to grasp. In situations like this, people rely on trusted advisers or opinion leaders who in turn rely on trusted advisors to frame the situation for them. Their advice can make one approach seem to be not just the best bet, but a moral imperative.

Bias of Optimism

Overall, we have a greater willingness to accept findings that would be welcome, if they were true. This pattern of thinking leads us to overestimate the chances that one thing or another will work out well for us. For example, people's ideas about what the stock market will do are typically rosier than past experience would suggest. The same goes for real estate prices.

According to Kahneman, three factors underlie erroneous optimism: we exaggerate our own skill; we overestimate the amount of control we have over the future, and we neglect to consider the possibly superior skills of others.[19]

We also habitually underestimate the odds of failure in complex systems. Years ago, I was stunned when my colleague William J. Broad reported that engineers had concluded the odds of catastrophic failure in any given mission of the Space Shuttle were about 1 in 70. The number seemed way too high. But in 135 shuttle launchings there have been two catastrophic failures—the explosion on launching of *Challenger* in 1986 and disintegration on descent of the *Columbia* in 2003. In other words, about 1 in 70.

Aversion to Loss

We fear loss much more than we desire gain. Kahneman and Tversky attributed this to our Paleolithic past, when survival was hardly a sure thing. In that kind of subsistence existence, they reasoned, a benefit is nice, but a loss can be catastrophic. Perhaps this is one reason why people who are asked to choose between $100 now and $120 a month from now will almost always take the money and run.

Aversion to loss also encourages us in the so-called *sunk cost fallacy,* in which we persist in unprofitable or even obviously doomed endeavors because of the time, effort, and money we have already put into them. (My family calls this practice "pouring good money after bad.") A reasonable person would look ahead, calculate the costs and benefits going forward and make a decision on that basis. As for the rest of us . . .

INATTENTIONAL BLINDNESS

A few years ago, I attended a presentation by Max Bazerman, a professor at Harvard Business School. He was talking about what he calls "predictable surprises"—bad news we have every reason to anticipate but, for various reasons, do not. To begin his presentation, he showed his audience a short film. He told us the film would show a group of young people, some in white shirts and some in black, passing basketballs to each other. He asked us to count the number of times someone in a white shirt threw the ball.[20] (If you want to view the film before you read the spoiler, read no further and search on "selective attention test" on YouTube.)

Though the counting task was harder than it seemed, many of us got it right. But then Bazerman asked if any of us had noticed anything unusual in the film. In his audience of about 200 people, no one raised a hand. So he ran the film again. What none of us had noticed was that in the middle of the ball-tossing a "gorilla" walked through the players, stopped in the middle of the room, turned to face the camera, and thumped its chest.

It seems incredible that none of us had seen the person in the gorilla suit, but since then I have shown the film often, and not a single person has seen the gorilla. (In another version, a woman carrying an open umbrella walks through the players. No one notices her either.)

Researchers call this phenomenon *inattentional blindness*—our failure to see something that is right in front of us because our attention is engaged elsewhere—in this case, on counting the number of passes. For Bazerman, the film demonstrated the way close attention to a task can blind us to phenomena around us we really ought to pay attention to. He said later that when he played the video for a group of economists, one of them confronted him afterward, furious that he had been somehow been found wanting in failing to notice the gorilla, even though he had produced an accurate count of the number of passes. "You gave me a task and I completed it correctly," he fumed. Perhaps he was one of the many on Wall Street who by 2008 had paid so much attention to their bonuses they neglected to notice they had driven the economy into a ditch.

In life we need to pay attention. But often we don't.

DISASTER BY DESIGN
One recent—and heartbreaking—example of the predictable surprise, the disaster we almost design ourselves to encounter, was the Deepwater Horizon oil spill, which occurred in 2010 when an offshore oil rig exploded in the Gulf of Mexico, killing eleven workers. According to the presidential panel that assessed the disaster, it should have been obvious to people in the oil industry and the people who regulated it that industry safety standards and spill response preparations were not keeping pace as drilling in deep water increased. As William K. Reilly, former administrator of the Environmental Protection Agency, put it when the commission reported its findings, "In effect, our nation was entirely unprepared for an inevitable disaster."[21]

Federal policies that actually encourage people to build in areas vulnerable to coastal flooding are producing disasters by design every storm season. The same is true in what Roger Kennedy, former head of the National Parks Service, called "the flame zone," areas in the West managed to make fire more likely even as more and more people build homes there. Part of that problem, he wrote in his book *Wildfire and Americans,* was the fact that people moving into those areas could rely on government firefighting efforts.[22] It was more than tragic to him that people are killed every year fighting these fires, usually lower-income people who die to protect the homes of the well-off, who should have known better than to build in the flame zone in the first place.

CULTURAL COGNITION

When the physicist Robert Park says that most people "judge science by how well it agrees with the way they want the world to be," he is talking, more or less, about a phenomenon known as cultural cognition.

Dan Kahan, a researcher at Yale Law School, described it at a conference in Washington, D.C. in 2011 on the communication of risk to the lay public.[23] He noted that many questions we face as a society are answerable. For example, is human-induced climate change a problem? That is a question science can answer—and science has answered it: yes.

Yet a large segment of society rejects that answer.

According to Kahan and other researchers in this field, the reason is a seemingly powerful—though usually unconscious—desire to hold opinions that put us in sync with social aspirations we have for ourselves, the people we wish to be associated with, and so on. Researchers who study this phenomenon say it overpowers any amount of information we may receive. "People tend to conform perceptions of risk to visions of what an ideal society would be—visions they share with other people," Kahan said.

For example, people who value individual effort over collaborative, cooperative efforts may turn away from the facts of climate change—because effective action on climate change is almost certainly going to require national or even international regulation, the kind of thing people in this group would despise.

To cite another example, people who believe natural things are best may turn away from information suggesting that genetically engineered crops may be safe—and even better for the environment than conventionally raised crops.

When people are presented with information that challenges their world view, they will ignore it, interpret it erroneously, or dismiss it. That is why many climate dissidents presented with yet more evidence that the problem is real dismiss it as yet more proof that the conspiracy is larger than they had thought.

I first heard these ideas discussed in detail in the summer of 2010, when I was asked to address leaders of the National Academy of Sciences on the public's understanding of science. Kahan and Anthony Leiserowitz, director of the Yale Project on Climate Change, were on the program.[24] The examples they used then were attitudes toward climate, gun control, nuclear power, and nanotechnology.

Kahan said his interest in this research began with his assumption that people might have opinions, but if new information became available, "they would take in new information and if necessary adjust their views in light of it." But then, he said, he realized that "if things actually worked like this, there would be a progression to universal enlightenment"—something we have yet to witness.

In fact, he and Dr. Leiserowitz told the group, people assess new information in light of their prior perceptions. And they look for conforming information, dismissing facts that don't fit their views.

Social scientists have recognized this thinking pattern for decades, at least, but it is drawing new attention now, in particular, because of the widespread inability of climate scientists to understand why so many Americans are so resistant to their message about the threat of greenhouse gas emissions.

Kahan and Leiserowitz group people they study on two scales: egalitarian-hierarchical and communitarian-individualistic. They found that people with a communitarian/egalitarian frame of mind are prepared to accept information about environmental risks like climate change. The concerted action needed to avert it is not uncongenial to these people. But people who are hierarchical and who are individualistic rather than communitarian are much more likely to be skeptical.

Leiserowitz and Kahan found these factors were good predictors of people's views not just on climate but also on topics as varied as gun control, use of the vaccine against human papilloma virus (a major cause of cervical cancer), storage of nuclear waste, and the carrying of concealed weapons.

What does this tell us? Among other things, it suggests that merely exposing people to more information is not necessarily going to lead them to good decisions. As a journalist, I hate this idea, but the evidence for it is pretty convincing.

And it means that when you find yourself reflexively accepting or rejecting some assertion or other, it might be wise to examine your own patterns of thinking. Try this thought experiment: when you assess Donald Trump as a political leader, does his lurid personal life tell against him? How about Bill Clinton? Depending on your politics, you are probably readier to excuse one than the other.

STATISTICAL INCOMPETENCE

Meanwhile, there's another problem, in some ways the biggest of all. Americans, in aggregate, have a very poor understanding of statistics. And even if, like me, you know a little about the field, your intuitions can still lead you astray. I experienced this phenomenon a few years ago, when I was one of a hundred or more science journalists who participated in a workshop on medical evidence. It was organized by the Knight Center at the Massachusetts Institute of Technology, which supports efforts to improve science journalism.

One of the presenters was Josh Tenenbaum, a professor of cognitive science at the university. He began his presentation by announcing that he was going to determine who in his audience had extrasensory perception. He had a fair coin—a perfectly balanced coin—and he would flip it five times. Then he would transmit mind rays to us, telling us how it landed. We would absorb his mind rays, if we could, and write down whether each flip was heads or tails.

There are thirty-two possible combinations, but about a third of us, including me, wrote down HHTHT. Another quarter of us wrote the obverse, TTHTH. No one wrote down HHHHH, though the odds are identical to the odds of HHTHT—if a coin is evenly balanced it ought to turn up five heads in a row, or five tails, about 3 percent of the time.

This and similar tests tell us that most people know what randomness looks like. Even among "sophisticated populations," as Tenenbaum called them—and probably he would have classed a room full of science journalists as a sophisticated population—people gravitate toward the same guesses. Why should we care about this little bit of cognitive arcana? Because it comes into play whenever anyone talks about cancer clusters or "outbreaks" of autism or the like.

Five heads in a row does not *feel* random. The sequence does not match our intuitive sense of what randomness looks like. We don't understand that if something—like cancer—is widespread and distributed more or less randomly over a large population, there will inevitably be clusters of it here and there. If there are no clusters, a statistician could tell us, then the distribution is too even to be random.

But three cancer cases in the same neighborhood or three children with autism in the same school look alarming. "Aha," we think, there must be "something in the water." And sometimes there is. But usually there is not. There is only the reality of statistics.

REGRESSION TO THE MEAN

Imagine you are the boss and you have two employees. One of them does a task unusually well and you praise her. But then her spectacularly good day is followed by days of ordinary performance. Meanwhile, the other employee does a task unusually badly and you berate her. Her bad day is followed by ordinary—i.e., better—work.

As a boss, you may conclude that praise is an ineffective management tool, that punishment works much better. Is that right? No. What you have just experienced is called regression to the mean, the tendency of whatever it is you are measuring to return after an extraordinary episode to ordinary levels.

According to Kahneman and Tversky (and many others), we do not expect to see this regression even when it is bound to occur, and when we see it occurring we invent spurious reasons to explain it—for example, that praise encourages workers to slack off.

THE WISDOM OF BAYES

Few people have ever heard of Thomas Bayes, an eighteenth-century English statistician and Presbyterian minister who wrote, among other things, "An Essay towards Solving a Problem in the Doctrine of Chances." The work relates to something statisticians call *conditional probability*; that is, the probability of something, given something else. For example, what is the probability that a card is a king, given that it is a face card? Since there are twelve face cards in a deck and four of them are kings, the probability that a card is a king, given that it is a face card, is 4/12 or one-third.

Conditional probability arises more commonly than you might expect. For example, assume there is a test that will tell you, with 98 percent accuracy, if you have cancer. That is, if you have cancer, the test will be positive 98 percent of the time, and if you don't have cancer, it will be negative 98 percent of the time.

Then imagine that you have had the test and the result was positive. Should you worry? You might if you did not consider this important question: how common is cancer? In other words, you must be sensitive to what statisticians call the "prior probability" of outcomes.

In this example, suppose we know that at any given time .5 percent of people—one half of one percent—actually have cancer (the prior probability). Then imagine 10,000 people take the test.

Given cancer's prevalence in this example, fifty of them will have cancer. If the test is 98 percent accurate, forty-nine of them will receive a positive result, a cancer diagnosis.

Meanwhile, the other 9,950 people are also getting their test results. These people don't have cancer, but because the test is only 98 percent accurate, 2 percent of them—199 people—will receive a positive diagnosis.

In other words, of the 248 positive results, 199 are false positives—people receiving a diagnosis of a disease they do not have. In this example, only about 20 percent of the people who receive a positive cancer test result actually have the disease.

This example was offered by John Allen Paulos, a mathematician at Temple University, who writes about it in his book *Innumeracy*.[25] "This unexpected figure for a test that we assume to be

98 percent accurate should give legislators pause when they contemplate instituting mandatory or widespread testing for drugs or AIDS or whatever," he wrote. And, as he notes, many tests in wide use are even less reliable.

How Do You Know?
Sometimes our most erroneous thinking arises because of ideas we don't think to question because they are so obviously correct.

Though some of my neighbors disagree with me, I think an example of this kind of thing played out recently where I live, on Chappaquiddick Island, off Martha's Vineyard, in Massachusetts. The island has only one paved road, and in July and August many people bike along it, including many children. A number of people thought it was obvious that things would be safer if we had a bike path. Until I looked into it, I thought so too.

But it turned out to be far from clear that the proposed remedy—a two-way bike path along one side of the road—would make things better. In fact, some evidence suggests that this kind of bike path actually makes things worse. The problem occurs when another street or even a driveway meets the road/path. Drivers typically look to the left for an oncoming car; they forget that a bicycle might be coming from the right.

Another proposed remedy, creating a bike lane and marking it with a white line along the side of the road, also sounded good to me, until I discovered evidence that when bikers and drivers share the road, drivers tend to slow and give bikers lots of room. Where there is a white line, motorists drive as if every inch of macadam up to the line belongs to them. Result? Cars drive much closer to bikes than they would otherwise.

I don't know what research will ultimately tell us about this question, but the idea that a bike path could make things more dangerous, not safer, is highly counterintuitive. Still, it is frustrating when people who assert something is necessary cannot say why, except that "it's obvious."

Donald E. Shelton, a judge in Ann Arbor, Michigan, encountered something similar when he began looking into the idea that people will expect more of real-life detectives if they watch television programs in which fictional detectives use fingerprints, DNA tests, ballistic analysis and other high-tech tools. Some judges and prosecutors call it "the CSI effect," after television programs about crime scene investigators.

"Many attorneys, judges, and journalists have claimed that watching television programs like CSI has caused jurors to wrongfully acquit guilty defendants when no scientific evidence has been presented," Shelton wrote in *NIJ Journal,* a publication of the U.S. Department of Justice.[26] The "CSI effect" is worrisome, some say, in that it presses law enforcement agencies to make unneeded commitments of time, money, and equipment in investigating cases.

Alternately, prosecutors who anticipate presenting relatively little scientific evidence may preemptively remove fans of the programs from their juries, assuming they will interpret an absence of such evidence as an indication that the prosecution's case is weak.

But when Shelton, a member of the faculty of Eastern Michigan University, teamed up with two criminology professors at the university to actually survey jurors, the association did not hold up. Jurors' responses to scientific evidence seemed to have no relation to their television habits.

"Although CSI viewers had higher expectations for scientific evidence than non-CSI viewers, these expectations had little, if any, bearing on the respondents' propensity to convict," Judge Shelton wrote.

He added, though, "many laypeople know—or think they know—more about science and technology from what they have learned through the media than what they learned in school. It is those people who sit on juries."

Where does all this leave us? Well, when you are contemplating a science or engineering question, think not only of the technical issues but also of the way you are doing your thinking. When you look into something, embrace what the information theorist Jonathan G. Koomey calls the "scientific outlook."[27] That is, be willing to accept what you find, and be willing to discover that you are wrong.

Be self-aware enough to know what kinds of ideas are likely to skew your views, to cause us to accept or reject information unthinkingly. Consider these questions: Where do you get your information? Whom do you trust, and why? With whom do you normally agree? Do you agree all the time, and do you trust both the values and the judgment of those people? To whom do you give the benefit of the doubt? And finally, what do you find completely outrageous, and why?

I am not suggesting that everyone needs to take a course in statistics, but it would be great if everyone did. It would also be great if the people who design college entrance exams emphasized questions about statistics—the move would force the nation's high schools to spend more time teaching the subject.

But failing all that, we need to keep asking a simple question: How do we know?

Thinking about Risk

A while ago, the Environmental Protection Agency surveyed a group of Americans about the environmental problems they worried about. Then they asked a sampling of scientists the same question. In each group, there were a number of issues that provoked widespread worry. The problem was, overall, the groups were not afraid of the same things. In fact, when the EPA researchers compiled lists of the ten things each group worried about most, there was hardly any overlap.

The public was worrying about oil spills, hazardous waste dumps, releases of radioactivity, and pesticides and other chemicals in the environment. The scientists saw all these as risks, but understood that they had relatively limited and short-lived effects, so they put them at the bottom of their lists. Though they worried a bit about pesticides, for example, their concern was for the people who manufactured or applied them in bulk, not for people who might consume residues on their produce. What was alarming the scientists? Global warming, habitat loss, and loss of biological diversity.

In general, the scientists had chosen the more significant concerns. Alarmist reports notwithstanding, pesticide residues in the food supply are not a threat to public health. People who eat fruits and vegetables, even those labeled "organic," consume far more pesticides than they realize, in the form of natural compounds plants make to protect themselves from insects and other threats. Many of these "natural" pesticides work just the way the synthetics do. And there is no doubt eating your vegetables is good for you.

Oil spills do not constitute a major global hazard, either. Even the Deepwater Horizon disaster in the Gulf of Mexico does not seem to have left the lasting damage some had predicted. Anyway, if you worry about oil in the ocean, worry about ordinary day-to-day leakage from ships. Far more oil seeps into the oceans that way than through attention-getting spills.

That is not to say we should soak our fields in pesticides or spill oil. Chemical pesticides interfere with the environment, and organic farming methods are undoubtedly better for the land and possibly even more productive—even if they aren't better for our health than conventional crops. And oil spills help no one except the companies hired to clean them up and the environmental groups that draw new members and donations when one of them occurs. But how prominent should these problems be when we set our environmental priorities? The question is important because while we can afford to spend money to protect the environment, our budget for the task is not infinite. We must spend it wisely to thwart threats that could change our environment for the worse, permanently. Instead, we tend to go by the public's list, spending vast amounts of money on situations that pose little hazard and neglecting issues whose threat is vast.

People in industrialized countries live safer lives than ever before, yet we worry more about risk. Researchers who study risk and people's perception of risk offer several explanations.

First, the hazards we face are quite different from those our ancestors confronted. Technology has greatly increased humanity's ability to control our environment, and we are far more dependent on technology than people were even a generation ago. But the powerful, complex technologies on which we depend have powerful, complex—and sometimes unwelcome—effects.

Second, we have more information about the risks confronting us. Some of this information is imperfect, specious, or even fraudulent, but that does not make it any less worrying.

And so, we worry more. We are a loss-averse species and we have more to lose.

Finally, we have lost trust in government, religious authorities, business leaders, and social institutions which once might have been able to reassure us. Trust in Congress is at a low ebb—but trust in almost everyone seems to be at a low ebb as well.

Meanwhile, the debate over what is risky and what is not has become thoroughly politicized. Experts on risk perception say that is the inevitable consequence of government efforts to regulate our exposure to hazards. They say that once you establish a regulatory agency, give it a jurisdiction, and tell it to make rules, it will start doing research on what and how to regulate. In doing so, it will uncover new situations demanding their own new rules, leading to more findings, more rules, and so on.

This rule-making does not occur in a vacuum. Regulation becomes a forum for competing interests. Scientific arguments become proxies for political or economic or philosophical or even religious disagreements. In the end, even what looks like a purely scientific decision for policy makers eventually turns out to hang on the answer to this value-laden inquiry: How do we want the world to be?

Over the years, a whole field of social science research has grown up around figuring out what we are afraid of, and why. Today there is wide agreement on a number of points.

- We fear the uncontrollable more than things we can control. This is why we fear flying far more than driving, though driving is far more dangerous.

- We fear things imbued with dread. Although cancer is not always the unremitting horror it was in the days before adequate pain relief, we still dread it far more than we fear diseases like heart failure, which in many ways can be at least as miserable and kills more of us than cancer does.
- We fear the catastrophe more than the chronic condition, even if the chronic condition carries the same risk or worse.
- We fear things imposed on us—water pollution, say—more than things we expose ourselves to voluntarily, like fatty foods or cigarettes. If we cannot tell whether or not we have been exposed, we are even more afraid.
- Things with delayed effects are more frightening than things whose effects are immediate.
- New risks are scarier than old risks.
- If we don't trust the person or agency telling us about the risk, we are more afraid.
- A hazard with identifiable victims is more frightening than one whose risk is spread over a large population. That is why mad cow disease is more frightening than a diet rich in red meat, even though mad cow disease has killed practically nobody and heart disease—worsened by a beefy diet—is a leading cause of death.
- We worry more about things that affect future generations than threats to ourselves alone.
- We fear things we cannot see, like radiation, more than things we can see, like sooty pollution from coal-fired power plants. So we fear nuclear power plants, which in the United States have killed no one, far more than we fear pollution from coal-fired power plants, which causes lung problems that send tens of thousands of people to the hospital every year in the United States and, by some estimates, kills 5,000 of them or more annually.
- We are more afraid of things that are artificial, synthetic, or otherwise human-made than we are of things that

occur naturally. For example, some people worry a lot
about the presence in the environment of synthetic chemicals that mimic the actions of hormones like estrogen.
They don't worry at all about the estrogenic effects of a
diet rich in soy.

In fact, a good way to make people afraid of something is to
identify it as "chemical." For example, if I offered you a glass of dihydrogen monoxide, would you drink it? Maybe not, unless you knew
(or deduced) that dihydrogen (H_2) monoxide (O) is H_2O—water.
The natural world is made of chemicals—they are the stuff of life.

Politicians, corporations, or others with an axe to grind capitalize
on our widespread inability to react sensibly to risk. As a result, we
may waste money or issue needless regulations to protect ourselves
against trivial or even nonexistent threats. Or they may pose a
problem, but not for the reasons we think. Or they may be problems whose cure may make matters worse.

PROBLEMS THAT DON'T EXIST

In 1997, a study appeared in the journal *Pediatrics* with an alarming
finding. More than 225 clinicians who evaluated more than
17,000 girls aged 3 to 12 in the course of normal physical exams
found that many were displaying signs of puberty "at younger ages
than currently used norms."[28] The researchers suggested that
chemicals in the environment, particularly chemicals that mimic
the action of the female hormone estrogen, should be investigated
as a possible cause of the situation.

The study received abundant attention in the press. Within
months, many people knew that this change in puberty had occurred among American girls and believed that synthetic chemicals in the environment were almost certainly the cause. But few
people read the actual study. If they had, they might have asked
whether its data were powerful enough to support its frightening
conclusion.

Later, critics of the study noted that the clinicians in the study
were asked to assess how many of their preteen girl patients were

developing budding breasts, a characteristic the researchers had defined as their marker of puberty. But because the clinicians did not actually touch the girls, it was difficult for them to differentiate between fatty tissue and breast tissue. What they were observing was almost certainly that more of their preteen girl patients were overweight.

Anyway, while it is difficult to tell exactly when a young girl has entered puberty, one part of the process is easily detectable and has been tracked for decades: age of first menstruation. And, according to many medical authorities, the age of menstruation in the United States is unchanged since the 1950s at least.[29] American girls begin menstruating at younger ages, overall, than they did in the nineteenth century, and sooner than girls in many poor countries of the world, but those differences are the result of better nutrition. Overweight girls may begin menstruating somewhat earlier. As the Committee on Adolescence of the American Academy of Pediatrics put it, a higher gain in body mass index (BMI) during childhood is related to earlier menstruation. Even so, the academy said, American girls "are not gaining reproductive potential earlier than in the past."[30]

Since then there have been a number of other reports, and new evidence may change expert opinion. But so far the best—some would say the only—hard evidence we have about the onset of puberty among American girls suggests that nothing much has changed—except that American girls are getting fatter.

Of course, there is such a thing as abnormally early puberty, which can be caused by conditions like brain tumors. Many experts worry now that wide acceptance of the idea that chemicals in the environment are causing girls to enter puberty earlier may cause some of these cases to be ignored.

Activists making the puberty-chemical link cited the decline of amphibian populations around the world as more evidence for their case. The amphibians—frogs, toads, salamanders, and other creatures—are declining all over, and in many places researchers were reporting the appearance of amphibians with ambiguous sexual organs, or multiple sets of organs, too many legs, or other severe deformities. More evidence of the evil effects of hormone mimics in the environment is mounting, advocates asserted.

Later, though, it turned out that while environmental pollution may be at play, the major factors in amphibian decline are climate change, habitat loss, and a bizarre kind of parasitic infection. In this case, there was definitely a problem—vanishing and deformed amphibians—but the explanation put forward was not adequate to account for it.

According to the U.S. Geological Survey, tests of water in many rivers in the United States turn up signs of a variety of drugs and other chemicals. But so far at least it is impossible to link this situation to any real problems. Is that to say it's not something to watch? Of course not, particularly since chemicals approved for use are usually tested one by one, but as water moves through sewerage systems, it can pick up chemicals that have combined in vast and potentially complex ways.

Low sperm count, supposedly caused by exposure to these chemicals or other factors, is another problem that, so far anyway, does not seem to exist. Though there are researchers who assert their studies find it, the overwhelming weight of evidence goes the other way. Nevertheless, the idea has drawn wide attention among activists. Criticism of the data and study methods soon erupted but was less noticed, even as other researchers produced contradictory findings. In 1999, an expert panel convened by the National Academy of Sciences said it could not reach consensus on the issue because the research was so flawed. Other research suggests the problem does not exist.[31] True, there has been growth in the United States in the percentage of couples who have difficulty conceiving, but for most of them the issue is maternal age—women are putting off childbearing. According to the U.S. Census Department, the age of first marriage in the United States has risen from just over 20 in 1950 to almost 27 in 2010.[32]

The supposed "breast cancer epidemic" on Long Island was a similar problem that did not exist. Certainly women on Long Island were being diagnosed with breast cancer, and at higher rates than nationally—higher even than women in nearby areas. It turns out, though, that what Long Island has in relative abundance is not breast cancer but rather affluent women with access to good insurance coverage. When researchers looked at similarly affluent areas (like Bergen County, New Jersey, or Marin County, California), they found similar "epidemics."

Fortunately, alarm over nonexistent problems is unusual. But when the alarm bell is ringing, it is worth stopping to ask, "how do we know anything untoward is really going on?"

Problems, but Not for the Reasons We Think

In 2011, the International Agency for Research on Cancer, an arm of the World Health Organization, declared that it was "possible" that cell phone use could cause brain cancer. The panel cited what appeared to be an association between cell phone use and small increases in certain rare cancers. The idea was that the radiation emitted by the phones damages DNA in cells in the head, leading to malignancy.

But there are problems with this theory. First, even as cell phone use explodes, brain cancer rates in the United States are more or less unchanged. According to the National Cancer Institute, between 1975 and 2011 the annual rate of diagnosis has been bouncing around between 5.8 and 7.1 cases per 100,000 people. It was 6.4 in 2013.[33] Meanwhile, research attempting to find a connection has produced results that are at best equivocal, at worst (possibly) fraudulent.[34]

Cell phones do not produce the kind of radiation—ionizing radiation—that can alter the genes inside a cell and give rise to cancer. As researchers put it in the *Journal of the National Cancer Institute,* in reporting results of a study in Scandinavia and Switzerland, "the lack of an exposure-response relationship, given our finding that risk was related to neither the amount of mobile phone use and [*sic*] nor the location of the tumor, does not support a causal interpretation."[35]

In other words, there is no reason to think cell phone use causes brain cancer.

Of course, absence of evidence does not equal evidence of absence—the fact that we cannot see evidence of a problem does not necessarily mean there is no evidence. But as the editors of the cancer journal pointed out in an editorial, "In considering the need for future cell phone health research, it should be kept in mind that in addition to the negative epidemiological data"—i.e., the absence of any evidence that cell phone users are more prone to brain tumors than nonusers—"there is no known biologically plausible mechanism by which nonionizing radio waves of low energy can disrupt DNA and lead to cancer. The photoelectric effect is not a matter of opinion; radio frequency energy absorption cannot break DNA molecules and carcinogenicity studies in animals are rather consistent in showing no increases in cancer following radio frequency energy absorption."[36]

Or as the oncologist Siddhartha Mukherjee put it, "If cellphone radiation is causing cancer, it is doing so through a mechanism that defies our current understanding of cancer genetics."[37] In an essay in the *New York Times,* Dr. Mukherjee, author of the acclaimed "biography" of cancer, *The Emperor of All Maladies,* noted as an aside that the WHO's list of "possible" carcinogens also includes pickles. But we are not afraid of pickles.

The *Times* itself is not immune. One study—involving only forty-seven people—found that "less than an hour of cell-phone use can speed up brain activity in the area closest to the phone antenna, raising new questions about the health effects of low levels of radiation emitted from cellphones," the newspaper reported.[38] The story noted that "a few observational studies have suggested a link between heavy cellphone use and rare brain tumors, but the bulk of the available scientific evidence shows no added risk. Major medical groups have said that cellphones are safe . . ."

Still, soon after, the newspaper's readers were treated to a litany of cell phone don'ts: don't hold the phone up to your ear (turn on the speaker), don't carry it in a pocket (keep it in a purse or briefcase or belt clip "that orients it away from your body"), don't use it if service is iffy (that requires more oomph from the phone and therefore, presumably, more radiation), and don't allow children to use cell phones for extended periods, or at all.[39]

Brain cancer is definitely a problem. But given that rates of brain cancer are more or less steady and given that it is physically impossible for the radiation emitted by a cell phone to cause cancer, why does the idea that cell phones cause cancer refuse to die? Maybe it has to do with the emergence of companies that market devices that purport to shield users from radiation, or even remove it from the air. That's why I don't expect the furor to die down, even though study after study reports, as one did in 2011, "no increased risk of brain tumors with long-term use" of cell phones.[40]

The cartoonist Nicole Hollander summed all this up in a cartoon in 2008 in which two people are conversing. "In fact there is no known biological mechanism for the phones' nonionizing radiation to cause cancer," the first says. "Easy for you to say," the second replies. "The truth is out there."[41]

In May 2016, another study—of rats—purported to demonstrate a potential—but very small—additional risk of brain cancer. The findings generated public unease, but the Associated Press (and some other news organizations) greeted the news with appropriate skepticism. Among other things, the agency's report on the findings noted that the study "had enough strange findings that it has caused other federal scientists to highlight flaws in the research."[42] In the study, financed by the National Institutes of Health, rats were subjected to cell phone radiation for nine hours a day from the womb to age two. Two to three percent of male rats developed tumors but, strangely, females did not seem to be affected at all. Also, rats not exposed to the radiation died at twice the rate of rats that were. That aspect of the study did not receive a lot of attention. Meanwhile, rates of brain cancer in humans remain stable.

Perhaps it was inevitable that people would attempt to enact legislation to protect us against the cell-phone menace. Perhaps it was also inevitable that those people would be members of the Berkeley,

California City Council. In 2015 they passed a "Right to Know" ordinance—"not actually backed by science," as the *New York Times* reported, requiring cell-phone sellers to tell customers that the devices could be hazardous to their health.[43]

Given its complete lack of support from the scientific community, the ordinance has come under criticism, not least from makers of cell phones. But it had its defenders, most of whom cite not scientific evidence but rather the importance to the First Amendment of allowing people to speak freely about the issue. Inevitably, there were lawsuits, some of which have been consolidated into a class action suit against Motorola and other companies connected to the cell phone business.[44]

Though they will not give you brain cancer, cell phones can still be killers. Texting while driving is now approaching drunkenness as a cause of death on the road, and many of us (including myself) have known people who have died or been severely injured in a crash involving a driver using a cell phone. Nevertheless, most of us are not afraid of texting, even while driving.

The most notorious example of a problem widely attributed to an erroneous cause is the supposed link between autism and childhood vaccination.

Rates of autism diagnosis have long been on the rise. According to the Centers for Disease Control and Prevention, early epidemiological studies in Europe in the 1960s and 1970s put the rate at about 1 in 2,500. By 2010, the rate was 1 in 68 American children.[45] Parents of children with autism noticed that reports of the condition began to climb in recent decades, since the advent of the combination measles-mumps-rubella vaccine, which came into wide use in the 1970s. And they noticed that their autistic children received the diagnosis shortly after receiving the vaccine. It could not be a coincidence, these groups reasoned. They theorized that a chemical called thimerosal, a mercury-based preservative used in vaccines, was damaging children's brains. As a result, thousands of parents refused to authorize the vaccination of their children.

But there is no evidence linking vaccines to autism.

For a while, advocates of the vaccine-autism link pointed to a study reported in the British journal the *Lancet* in 1998.[46] The author, Andrew Wakefield, a British physician, claimed that in his study—of only about a dozen children!—he discerned a link between the measles-mumps-rubella (MMR) vaccine and autism. The idea never garnered additional support among researchers; in fact, paper after paper debunked it. Also, vaccine makers abandoned mercury-based preservatives in vaccines, yet autism diagnosis rates continued to climb.

Finally, in February 2010, the *Lancet* retracted the paper, calling it a fraud, and the British government lifted Wakefield's license to practice medicine. But there are still plenty of people who believe the vaccine-autism idea. Many of them are depriving their children of needed vaccinations. For a while, that did not matter too much, at least in the United States. Because so many other children were being vaccinated, diseases like measles did not have much chance to take hold in our population. Epidemiologists call this "herd immunity."

But in places like Boulder, Colorado, and the Bay Area of California, where many parents joined the antivaccination movement, outbreaks of whooping cough and measles began to appear. In June 2014, when more than 800 cases of whooping cough appeared in the space of two weeks, California declared a whooping cough epidemic. By then, there had been 3,458 reported cases since the beginning of the year.[47] Health officials attributed the outbreak to large numbers of unvaccinated people living in affected areas.

Later, an epidemic of measles erupted in California, apparently spread among unvaccinated children visiting Disney World.

Over the years there have been many explanations offered for autism, beginning with distant parenting (the so-called "refrigerator mother" theory that made a splash in the 1950s and 1960s). The most promising ideas at the moment are that the condition is caused by some kind of genetic influence (which would explain why the condition is more common in some families) or fetal exposure to hormones, especially testosterone, in the womb, which might explain why the condition is more commonly diagnosed in boys than in girls.

Or the explanation might be more simple. According to the Centers for Disease Control and Prevention, studies suggest "that much of the recent prevalence increase is likely attributable to extrinsic factors such as improved awareness and recognition and changes in diagnostic practice or service availability."[48] It adds that tracking the prevalence of autism spectrum disorder "poses unique challenges because of its complex nature, lack of diagnostic biomarkers, and changing diagnostic criteria."

Some of it is undoubtedly due to the greater availability of educational services for children with the condition. Some of it is probably attributable to a decline in diagnoses of retardation in favor of autism diagnoses. Some of it may also be due to the wider acceptance of a category of autism called Asperger's Syndrome. With our increasing proclivity to medicalize every variety of the human condition, children who once might have been called "eccentric" now have a medical diagnosis.

Or maybe there really is "something in the water." But whatever it is, it is not mercury preservatives or anything else in vaccines. Pursuing that baseless theory wastes money that could be better used elsewhere. Nevertheless, the antivaccine movement has some prominent supporters. Among them are Robert Kennedy Jr., the son of the murdered senator, and the actor Robert De Niro. In 2016, the Tribeca Film Festival, which De Niro co-founded, announced that it would show a documentary, *Vaxxed: From Cover-Up to Catastrophe*, co-written by Wakeman. De Niro, the father of an autistic child, supported the plan to show the film, he said, adding that he and his wife, Grace Hightower, "believe it is critical that all of the issues surrounding the causes of autism be openly discussed."[49] And he added that the film was the first he had ever asked the festival to screen in the fifteen years of its existence. After an outcry, the festival showing was cancelled, but the documentary was screened elsewhere.

The online publication *Slate* took on the issue in its series describing events in the United States using, as the magazine puts it, "the tropes and tone normally employed by the American media to describe events in other countries."[50]

"Fears about Western medicine and distrust of the government are making it harder to combat" the nation's worst measles outbreak in years, the Slate article said. "Despite funding cuts that have impacted the country's byzantine and often insufficient health care infrastructure, vaccines against measles and other disease are widely available," it went on. "But in most regions of the country, they are optional, and many parents—under the influence of celebrities, political ideologues and radical clerics—choose not to have their children vaccinated . . ."

Some of the vaccine resisters are "religious extremists," the article said, but it added, "Political scientists say years of war, social disruption, and political scandal have left many Americans highly distrustful of authority, whether represented by the government, the media or health workers."

As a journalist, I was struck by the way the tone of this *Slate* piece matched the condescending tone of articles we produce about people we think are less sophisticated than we are. It was hilarious as applied to my fellow citizens—or it would have been, were the situation not so sad.

PROBLEMS THAT EXIST BUT ARE TRIVIAL

A few years ago, when I was science editor of the *New York Times,* the executive editor came to me with a story idea. Someone had just been bitten by a shark while swimming at a Cape Cod beach. He thought we should do a story about the menace of hungry sharks.

I told him that people's chances of being attacked by a shark are far lower than their chances of being struck by lightning. The real

shark issue, I told him, was the hunting of sharks for the production of shark fin soup, a delicacy in some Asian countries. This
hunting was pushing some shark species to the edge of extinction.
And because sharks are big predator species at the top of the food
chain, their loss would have tremendous negative impact. That's the
story we wrote.

But the headline "Shark Attack!" is a lot sexier than "An Endangered Species Fights to Survive." Even today, when increasing seal
populations on places like Cape Cod are attracting more sharks to
waters near bathing beaches, shark attacks are, basically, not worth
worrying about—at least not much!

Childhood allergies to nuts are a similar problem. A nonissue
just a few decades ago, these allergies have achieved media prominence. Are there some people with nut allergies? Yes. Are some of
these allergies serious? Yes. A young woman who went to my college died after going out with friends for a chili dinner; unknown
to her (and the waitress she asked), the chili contained ground-up
nuts.

But is the nut situation so serious that peanuts should be banned
from airplanes, lest their fumes somehow sicken vulnerable passen-
gers? Should schools be declared "nut-free zones" and school buses
cleaned ("I am tempted to say decontaminated," Nicholas A. Christakis of Harvard Medical School once wrote)[51] if a nut is spotted on
the floor?

Dr. Christakis said he was prompted to write his essay, which
appeared in the *British Medical Journal,* when, as he tells it, his
children's school went to what he regarded as ridiculous lengths to
protect students from nuts. Dr. Christakis, a professor of medical
sociology, acknowledges that nut allergies are real—about 3.3 million Americans are allergic to nuts, he wrote. And 6.9 million are
allergic to seafood. But, he said, only about 150 Americans die each
year from all food allergies combined. (Advocates for the allergic
put the number at about 200.) Others who study the issue say it is
far lower—in single digits. They say more people die in accidental
encounters with lawn mowers than from food allergies.

Christakis says it is helpful to view nut phobia as an example of mass psychogenic illness—epidemic hysteria. Also, he says, removing nuts from children's environment may actually contribute to nut sensitivity. And the more attention is brought to the issue, the more parents worry about it which, he said, "encourages more parents to have their children tested, thus detecting mild and meaningless 'allergies' to nuts. And this encourages still more avoidance of nuts, leading to still more sensitization." In outbreaks of epidemic hysteria, experts generally recommend calm, authoritative reassurance. That, Dr. Christakis says, is what is needed when it comes to nuts.

But that approach runs into trouble. A few months before Dr. Christakis's cri de coeur, *Harper's* magazine reproduced—and annotated—a brochure from the Food Allergy and Anaphylaxis Network, an advocacy group. The advertisement said that for some reason—"no one knows why"—more and more children are developing food allergies, including allergies to foods like nuts.

Like Dr. Christakis, the magazine conceded that food allergies are real, but it went on to say that "exaggerating the threat may actually do as much harm as the allergies themselves." It cited an increase in psychosomatic reactions to foods, increased anxiety among children thought to have these allergies, and even reduced expectations for what they can accomplish in life.[52]

The result was a storm of outraged correspondence. "I am eleven years old and I have been raising awareness about food allergies for as long as I can remember," one writer said. "You RUINED IT!" Another, identified as Robert Pacenza, executive director of the Food Allergy Initiative, conceded that "food allergy nonprofits have an obligation to ensure that young patients function well in society, but underestimating this health concern puts children at risk."

One correspondent, who reported a lifetime of severe nut allergies, took another view, writing that food allergies are "far rarer than news reports would indicate."[53] Ironically, researchers in Britain reported in 2015 in the *New England Journal of Medicine* that if infants deemed to be at risk of developing a peanut allergy were regularly fed foods containing peanuts, they were far less likely to have developed an allergy by age five.[54] The finding is important, the researchers said, because rates of peanut allergies have risen dramatically, for reasons that are not understood and sometimes hotly debated.[55] In the United States, pediatricians routinely advise parents not to let at-risk children eat peanuts until they are at least 3 years old.

According to the RAND think tank about a third of Americans believe they are allergic to one food or another. But when researchers from RAND studied scientific reports on food allergies published between 1988 and 2009—chiefly peanuts, tree nuts, cow's milk, hen's eggs, fish, and shellfish—they found the true allergy prevalence is closer to 1 or 2 percent. (These foods make up more than half of reported food allergies, the researchers said.)[56]

One reason may be that the simple tests for food allergies—measuring blood antibodies or pricking the skin, applying tiny amounts of the alleged allergen and looking for reactions—can give deceptive results. The so-called "gold-standard" of allergy tests, feeding patients minuscule amounts of the food in question, in disguised form, alternating with placebos, is far more time-consuming and expensive.

This lack of good criteria for diagnosis "limits our ability to determine the best practice for the condition," according to Paul Shekelle, director of the Southern California Evidence-Based Practice Center at RAND, the senior author of the study.

REAL PROBLEMS WHOSE CURE IS WORSE THAN THEY ARE

A few years ago, the federal government issued a recommendation advising people to limit the amount of fish they ate, because of fears that some species were contaminated with mercury. Later that week, I was at a meeting in Rhode Island also attended by an official of the state's Department of Health and we chatted about the issue.

He told me that in light of the new guidance his agency was about to issue its own advisory, warning Rhode Islanders not to eat fish caught in Narragansett Bay. I assumed he would be in favor of this step. But he was angry about it. After he explained, I could see why.

As is often the case, he said, the risk warning had come without adequate consideration of its context, in this case the characteristics of the people who eat fish caught in Narragansett Bay. On the whole, they are people with low incomes. For them, fish from the bay are a good source of high-quality, low-cost protein. If they stop eating this fish, will they start eating lamb chops instead? Unlikely. Probably they will turn to cheap fast food. Will it be more nutritious than the fish? No. Is the small, theoretical mercury risk worth the loss of such a supply of cheap, nutritious food? My friend in the Health Department thought not.

So this is a question to ask yourself whenever anyone proposes that something be done or not done or consumed or avoided or whatever: and then what? Suppose you don't eat fish you catch in Narragansett Bay. Is your diet and therefore your health likely to improve? If not, maybe this advice does not work for you.

David Ropeik, a former television journalist who writes and consults on risk issues, considered this topic in an essay in 2010 in the *Columbia Journalism Review*. He cited a study reported in 2007 in the British journal the *Lancet* on consumption of fish by pregnant women. The study found the benefits for fetal cognitive health conferred by the fats in the fish far outweigh any possible harm from the mercury. In other words, avoiding seafood does more harm than good.[57]

How did major news organizations around the world cover this finding? "They didn't!" Ropeik wrote. He cited this and other misguided reports on risks as evidence "of a larger trend—if-it-scares-it-airs." In other words, findings that something is potentially harmful are more likely to find the limelight than findings that something is not so bad. "Incomplete or imbalanced and alarmist information can lead directly to harmful decisions," he wrote.

Sure enough, in 2014 the Food and Drug Administration and the Environmental Protection Agency issued a report suggesting that pregnant and nursing mothers were not eating enough fish.[58] The agencies recommended fish low in mercury like salmon, trout, anchovies, and sardines but also said pollack, shrimp, tilapia, catfish, cod, and light canned tuna are safe. In fact, they suggested a minimum amount of fish pregnant or nursing mothers should eat—twelve ounces a week.

In reporting the change, the *Times* quoted Dr. Stephen Ostroff, the FDA's acting chief scientist, who cited studies showing that children born to women who consumed fish have higher IQs and better cognitive development than children born to women who do not. (Of course, that may be because people who eat fish are wealthier or otherwise living in ways more conducive to higher IQ test scores).

"A large percentage of women are simply not eating enough fish," Dr. Ostroff said, "and as a result they are not getting the developmental and health benefits that fish can provide." He noted, by the way, that the agencies do not believe fish oil supplements confer the same benefits as eating the real thing.

In the last fifty years or so, there has been a major increase in the use of synthetics like plastic and vinyl in products and processes that touch almost all aspects of our lives. In particular, these substances have found wide use in medical instruments, devices, and products. So it was alarming when studies began to suggest that exposure to some of the chemicals that leach from these plastics might be harmful.

One of the chemicals was DEHP, a plasticizer used in vinyl. Millions of people have been exposed to this chemical, either because they needed short-term treatment for one thing or another, or because they have chronic conditions requiring long-term treatment. Federal regulators say many of these procedures can leave patients exposed to elevated levels of the chemical. By some measures there have been five to seven billion days of chronic exposure to plasticized medical products, and one to two billion patient days of chronic exposure.

Advocacy groups say this situation is dangerous. They say studies show that rodents fed large amounts of the chemical can suffer adverse effects.

The Obama administration got into this issue in May 2010, when it published an online report by the President's Cancer Panel, saying

the proportion of cancer cases caused by environment exposures—
things like pesticides, industrial chemicals, plastic food containers,
and so on—had been "grossly underestimated."[59]

Advocates for people who claim they have been injured or sick-
ened by exposure to one of these substances typically say that people's
genetic make-up varies (undoubtedly true) and that something about
some people makes them vulnerable. This may also be true, but at
the moment it is unproved. The idea relies on correlation-implies-
causality. In this as with all correlation assertions, it is important to
remember that the correlation is only a prompt for further research. It is not the result of research. It is a question, not an answer.

Anyway, according to the American Cancer Society only about
6 percent of cancers in the United States are owing to exposures to
these substances or things like sunshine, a major cause of skin
cancer. The major cancer villains, it said in an online statement, are
tobacco—which causes 30 percent of cancer deaths in the United
States each year—and poor nutrition, obesity, and lack of exercise.
But a report saying we need to eat more vegetables and exercise
more is not the kind of report that attracts attention. We are not
afraid of eating too few carrots.

Researchers reported in 2015 that the most common cause of
cancer is probably bad luck.[60] The researchers theorize that the
changes in DNA that give rise to cancer occur most often when cells
divide. Cells in some kinds of tissue divide much more rapidly than
others. Rapid cell division, they said, explains why some kinds of
tissues are much more likely to develop malignancies. They said
that while a third of cancers could be attributed to inherited pre-
dispositions or environmental exposures (like tobacco), "the majority is due to 'bad luck,' that is, random mutations arising during
DNA replication in normal, noncancerous stem cells."

Are they right? We'll see. If you keep reading, you will discover
that in science knowledge comes not from this stunning finding or
that, but rather from painstaking replication of research work. Still,
I was glad to read this report. A few years before, I had been diagnosed with breast cancer. Well-meaning acquaintances often asked

me, their faces long with concern, why I thought I had been stricken. My reply was always the same: "bad luck."

In early 2016, the International Agency for Research on Cancer caused another furor by announcing that red and processed meats should be considered possible causes of cancer in people. Critics of the agency criticized its methods and some even accused its staffers of conducting their work with particular outcomes in mind. Eventually, we may learn more about those accusations. Right now, though, I am noticing one tiny fact amid the yelling: the IARC has spent decades testing almost a thousand things "ranging from arsenic to hairdressing" and has identified only one substance it is comfortable dismissing as NOT a probable human carcinogen. It's an ingredient in nylon used in toothbrushes and stretch pants.[61]

Some risks are so embedded in the public mind that it seems nothing will dislodge them.

Drinking during pregnancy is one. To be sure, women who drink a lot—who engage in what amounts to sustained binge drinking— risk giving birth to babies with the small heads, distorted features and other signs of fetal alcohol syndrome. Reasoning by analogy (I guess), many doctors now advise women who are pregnant or hoping to become pregnant to avoid alcohol entirely. Ethical and liability concerns would undoubtedly preclude putting this theory to the test of a controlled trial, so we will probably never know whether this caution is warranted. At the moment, there is little actual evidence to suggest that a glass of wine with dinner has any ill effects at all. In fact, it is a kind of folk remedy used in the third trimester of pregnancy to prevent premature labor.

Another risk many of us take as a given is the supposed connection between living near electric power lines and childhood leukemia. Whole books have been written about this problem, which does not exist.

Perhaps the most worrisome risk, to many people, is radiation. Nuclear energy is put forward as a possible energy source in a climate-challenged world, but it faces an intractable public relations problem: people's fear of radiation. This fear expresses itself in opposition to new plants, protests over the continuing operating of existing plants, and advocacy for radiation regulations so strict they may set exposure limits below natural background radiation we are all exposed to daily.

Fear of radiation has also made it impossible for food companies to institute large-scale irradiation of their products, one of the most effective and efficient methods of killing pathogens on food without damaging food quality.

And it is fear of radiation, more than radiation itself, that makes the so-called "dirty bomb" a true terror weapon. Exploding a bomb filled with radioactive cesium or some similar substances in a populated area might not produce enough radiation to hurt anyone, but people are so afraid of *any* radiation at all that the region would be rendered uninhabitable even if, in fact, it was safe.

Researchers studying the landscape around Chernobyl, the site in Ukraine of a disastrous 1986 nuclear reactor explosion, have discovered an abundance of wildlife, including elk, deer, and wild boar. People abandoned the so-called "exclusion zone" around the plant after the disaster and have not come back, but the animals are doing fine, according to scientists. They said the animals were as abundant in the Belarus sector of the zone as they are at four unaffected nature reserves in Belarus. And they said animal tracks were similarly abundant near the plant site, regardless of how heavily sites were contaminated by radiation at the time of the disaster.[62]

After a tsunami struck the coast of Japan in 2011, heavily damaging the nuclear facility at Fukushima, fears of radiation exposure prompted an immediate evacuation of a wide surrounding area. By some estimates, more than 1,000 people died from stress-induced health effects of this evacuation, many of them residents of nursing homes, patients in hospital intensive care units, or other vulnerable people.

So far, no one is known to have died from radiation from the plant failures. Most of it was swept out to sea, but it is estimated that even the most heavily irradiated areas would have exposed people to the equivalent of a yearly full body scan. That's not nothing, of course; many who criticize the use of these scans cite the radiation exposure they inflict.

But people are not afraid of scans.

Among other things, radiation fears have delayed establishment of a national repository for nuclear waste at Yucca Mountain, Nevada, and provoked a debate notable for the passion with which partisans, and the scientists they hire, argue diametrically opposed positions. As a result of that, the rest of us are left with temporary waste depositories—typically containers stored in pools of water—a far more dangerous situation.

Questions to Ask about Risk

Research on decision making and risk perception suggests that people are not pure economic actors making rational decisions about what is best for themselves, their families, or their societies. We are individuals acting as individuals, groping through uncertainty with Stone Age reasoning skills that leave us vulnerable to manipulation. As the decisions we are called upon to make drift further and further from our real life experiences, we are more and more vulnerable.

In 2011, I attended a conference in Washington, D.C. on what organizers called "the risk of risk perception"—the harms that occur when people's ideas about risk lead them to make decisions that are actually harmful to themselves or society.

As Baruch Fischhoff, a participant at the meeting and a leading risk researcher, wrote in an article about risk perception and health behavior, "misdirected communications can prompt wrong decisions, create confusion, provoke conflict and cause undue alarm or complacency. Indeed, poor communications can have a greater public health impact than the risks that they attempt to describe." His prescription: "it should be no more acceptable to release an untested communication than an untested drug."[63]

Until that rule comes into force, and I am not holding my breath, here are some things to think about when you think about risk-avoiding actions you might take as a person, or we might take as a society.

First, ask yourself what the real problem is. If you're worried about getting brain cancer from your cell phone, maybe you should think instead about the risk of driving while texting.

Next, what is the chance you'll be affected? You want to know the absolute risk as well as the relative risk. People who want to scare us typically give us relative risk figures. For example, something that raises your risk of heart attack by 30 percent (relative risk) sounds a lot scarier than something that raises your odds from 6 in 1,000 to 8 in 1,000 (absolute risk). By the way, that's roughly the excess heart attack risk for women who take hormone replacement, according to the Women's Health Study.

How long do you have to wait? A substance that raises your odds of illness thirty or forty years from now is not the same as something that is going to make you sick tomorrow. Nitrites in your cold cuts may be something to avoid, but they are not the menace *E. coli* in your spinach is right now.

Finally, if the risk relates to an exposure to some chemical or other, how long has the exposure been occurring? Many reports of newly discovered chemicals in drinking water are the result of improved sensing technology rather than increased or new exposures. The fact that something can be detected or measured does not mean that it is causing anyone a problem.

My colleague Andrew C. Revkin addressed this issue in his DotEarth blog after Tracy Woodruff of the University of California, San Francisco, published a report in *Environmental Health Perspectives* on "environmental chemicals in pregnant women," based on data from the National Report on Human Exposure to Environmental Chemicals, a broad federal survey of exposure to a range of chemicals. Woodruff described the findings as "surprising and concerning" and said it warranted "a systematic approach that includes proactive government policies." As Andy pointed out, the data "is focused on what is detectable, not whether the measured levels constitute a hazard either individually or cumulatively."[64] He noted that "despite modern societies being 'awash' in chemicals, childhood cancer rates aren't showing much change and overall mortality rates continue their long-term decline."[65] In the United States at least, and despite epidemics of diabetes and obesity, Americans are living longer and enjoying more years free of disability. (Those trends have altered recently, but we know why: white Americans are increasingly dying of opiate overdoses, a trend probably attributable to contracting economic prospects more than anything else.)

INVESTIGATING CLUSTERS

Often, people will discern what seems to be an unusual outbreak or cluster of one ailment or another. Sometimes they are real health emergencies. Often they are not.

Perhaps the most famous disease cluster in the annals of medicine is an outbreak of cholera in London in 1850. An alert physician who mapped the addresses of affected people eventually figured out what the problem was: they were all drawing water from a particular well, and it was contaminated.

Another cluster was the 1976 outbreak of a previously unseen variety of pneumonia among guests at the Bellevue-Stratford Hotel in Philadelphia, where an American Legion convention was underway. Today the disease is known as Legionnaire's disease. Epidemiologists tracked the source of the outbreak to bacteria growing in condensation in the hotel's ventilation system.

But it is important to be cautious when you confront something that looks like a disease cluster. Epidemiologists from the CDC routinely investigate clusters of conditions, often as mundane as diarrhea, to identify contaminated food or other problems. The process is relatively straightforward, except for one knotty problem: persuading the public the agency is taking the situation seriously without raising unnecessary alarm. The agency has issued guidelines for investigators.[66]

Clusters usually share several characteristics, the guidelines note. They involve particular health problems or outcomes ("cases").

Usually a particular cause or agent ("exposure") is suspected, and people usually think they know how the exposure produces the cases. Sometimes, though, when people report a cluster, there's no apparent cause. In that event, the CDC recommends a systematic approach, with several initial steps.

When a cluster is reported, investigators should gather information on the number of cases, the suspected exposure or exposures, the geographical area and time period in which they occurred, and the characteristics (age, sex, age at diagnosis, address, length of time living there, and so on). Investigators should note whether the health events are sudden onset (acute) or chronic.

Investigators should keep in mind that local authorities may not be using a good definition of "case" and, as a result, may be lumping unrelated health problems into the same category, creating the misleading appearance of a cluster. "A variety of diagnoses speaks against common origin," the guide puts it.[67] Also, some conditions are so common that apparent clusters are unlikely to be real clusters. For example, the agency's guide for cluster investigations says your lifetime risk of contracting some kind of cancer is one in three, so what seems to be a cancer cluster may be nothing of the kind. Because cancer risk rises with age, "cases among older persons are less likely to be true clusters." Birth defects are less common but the CDC says they occur in up to 2 percent of live births. Others put the figure still higher.

There may be no obvious exposure, or there may be too many. An investigation of a suspected disease cluster near a toxic waste site, for example, may disclose that it contains hundreds of chemicals. "An investigation of the site may indicate no immediate or obvious connection between exposure or disease,"[68] the CDC guide says. Also, because diseases like cancer take a long time to develop—they have a long latency period, in the jargon of epidemiology—investigators would look with suspicion on a supposed exposure-related cancer cluster among people who moved to a region only recently.

Over the years, researchers have developed a number of statistical screens to determine whether a cluster of cases is likely to be clinically significant. Each has its strengths and weaknesses. And of course absence of evidence it not the same as evidence of absence.

The recommendations, developed by state and provincial health officials from the United States and Canada, begin with some possibilities: what looks like an excess of cases is, statistically, not an excess at all; or that an excess exists, but there is no discernable cause-and-effect tie to any given exposure. Like case series reports, clusters may be useful for generating hypotheses, but they are unlikely to be useful in testing hypotheses. Nonetheless, community perceptions about a cluster may be as important or even more important than an actual cluster. As a result, the recommendations note, "achieving rapport with a concerned community is critical to a satisfactory outcome, and this rapport often depends on mutual understanding of the limitations and strengths of available methods."[69]

In short, regulatory agencies confronting public concern about disease clusters and possible toxic exposures must tread a narrow path. For example, in a fact sheet published in 2010, the National Toxicology Program, based at the National Institute of Environmental Health Sciences at the National Institutes of Health, says it began looking into possible health effects of Bisphenol A (BPA), a chemical used in the manufacture of polycarbonate plastics and epoxy resins.[70] The fact sheet notes that BPA is in wide use and that people can encounter it if it leaches from storage containers into food, or even if they handle certain types of cash register receipts. And it adds that the Centers for Disease Control and Prevention found detectable levels of BPA in 93 percent of the people it tested.

Although the toxicology program said there was "some concern" about BPA's effects on children, it added that there was "minimal" or "negligible" concern that it would be linked to other often-cited problems, like supposed early puberty in girls.

In 2014, the toxicology program put out a fact sheet on cell phones and brain cancer.[71] The program's staff is working with radiofrequency radiation experts, the fact sheet says, but it notes there is no evidence conclusively linking cell phone use "with any health problems." (If you are worried, the fact sheet advises, spend less time on your cell phone.)

2

THE RESEARCH ENTERPRISE

What Is Science?

The enterprise we think of as science has ancient roots. Some would say it began in the fourth century BC, with the Greek philosopher Aristotle. Others link the beginning of science, modern science anyway, to the inductive reasoning of Francis Bacon, known as "the father of empiricism," who lived in England in the sixteenth and seventeenth centuries. Or to Tycho Brahe, the Danish astronomer who in the sixteenth century challenged the idea that the heavens were immutable and organized a program of observation to prove it. Or maybe it was the French philosopher and mathematician Blaise Pascal.

And then there was Isaac Newton, the British physicist and mathematician who followed a century later and whose idea it was that physical laws that hold true on Earth hold true throughout the solar system—and that the world was one huge, related, and uniform machine, whose fundamental principles were known—or could be found out.

In any event the research enterprise that emerged from the eighteenth century looks to nature for answers to questions about the material world, tests those answers with observation and experimentation, and regards all answers as tentative, standing only until they are successfully challenged. We call this approach *methodological naturalism.*

The decision to limit science to what can be examined, observed, and tested in the natural world is arbitrary. (In a sense, therefore, creationists are correct when they say that science is its own kind of religion.) The state of Kansas became enmeshed in this issue in 2005 when, in the course of attempting to allow the teaching of creationism in public school science classes, it abandoned language in its science curriculum calling science the search for "natural explanations for what we observe in the world around us."

But since the Enlightenment, when the modern scientific age may be said to have begun, methodological naturalism has paid tremendous dividends in the form of useful knowledge. In other words, it works.

Differentiating science from nonscience is not always easy. For one thing, the term "scientist" did not come into use until the middle of the nineteenth century. Until then, practitioners were typically called "naturalists" or "philosophers." But there are some things that characterize science. First, science asks—and aims to answer—questions about the natural world, from the characteristics of subatomic particles to the structure of the universe to the operations of the mind. Science also tests answers to these questions with observation and experimentation, and relies on the data these tests produce, discarding ideas that do not hold up. In general, an idea not susceptible to testing is probably not a scientific idea. To put all this another way, for an idea to be scientific, it must have the capacity to be incorrect or, as the philosopher of science Karl Popper wrote, it must be "falsifiable."

That is why creationism or any other idea relying on the action of a supernatural entity is not a scientific idea. Some argue this requirement of testability also throws some of cosmology and particle physics out of the scientific tent, at least at the moment, on the grounds that it is impossible to test them without, say, constructing a particle accelerator the size of the universe.

There is no "truth" in science; all scientific knowledge is provisional, standing only until it is overthrown by new information. The most you can say about a scientific idea is that, so far, it has held up under testing. In fact, scientific findings typically generate a new round of questions, calling for new efforts to find new answers.

Answers to scientific questions are assessed by the scientific community as a whole through two processes. First, anyone seeking to publish findings in a reputable arena—a scholarly journal, in a talk at a professional meeting, or the like—must first submit the work to review by scientific peers. Second, other researchers must be able to replicate the findings. The processes of peer review and replication are what convince scientists that an idea may have something to it. A flashy finding, all by itself, is not enough.

Skepticism is the default position of science. Kenneth R. Miller, a biologist at Brown University, made this point to me when the educational authorities in Cobb County, Georgia, decided in 2005 that a textbook of which Miller was an author ought to carry a sticker on the front cover warning students that it contained material on evolution. "This material should be approached with an open mind, studied carefully, and critically considered," the sticker would say.

When I called Miller to ask him what he thought about the proposal, I expected him to be outraged. Instead, he said he thought the sticker was fine—but only as far as it went. Students should approach *everything* in the book that way, he said.[1]

I should not have been surprised. Science regards all answers as tentative. That may be why landmarks of science are not called the atomic facts or the germ facts or the facts about gravity, but rather the atomic theory, the germ theory, and the theory of gravitation. Or the theory of evolution.

In ordinary life, when we talk about something being a theory, we mean it is an idea that is not solidly proved, sufficiently supported, or otherwise fully cooked. The opposite is the case in science. In science, a theory is "a comprehensive explanation of some aspect of nature that is supported by a vast body of evidence," as Francisco Ayala, a Catholic priest turned evolutionary biologist, puts it in his book *Darwin's Gift*.[2] An idea has to acquire an extraordinary amount of support to achieve the status of theory, the pinnacle of scientific ideas. Theories are typically broadly applicable, intellectually coherent, and capable of generating testable predictions. An idea like evolution does not achieve theory status until it has survived challenge after challenge.

In fact, of all the theories of science, the theory of evolution is the best supported. But it is still subject to refinement and elaboration.

(A scientific idea not subject to that kind of challenge and alteration is intellectually dead.) So we cannot call evolution a fact. We
can say only that the theory of evolution has been subjected to unusual scrutiny for 150 years, and it has held up.

As Ayala writes, "many scientific theories are so well established
that no new evidence is likely to alter them substantially." For example, no new evidence will demonstrate that the Earth does not
move around the Sun (heliocentric theory), or that living things are
not made of cells (cell theory), that matter is not composed of atoms
(atomic theory), or that the surface of the Earth is not divided into
solid plates that have moved over geological timescales (the theory
of plate tectonics). In cases like these, when a scientific explanation
has been tested and confirmed time and again, "there is no longer
a compelling reason to keep testing it," Ayala wrote.[3]

That is not to say that someone won't one day find a loose thread
in a theory, tug on it, and unravel the whole thing. Consider physics.
By about the turn of the twentieth century, many physicists believed
their field had no new worlds to conquer; they had figured everything out. In fact, the story goes that the young Max Planck was
advised against a career in physics on the grounds that it would be
a professional dead end.

Within a few years Einsteinian relativity and Planckian quantum
mechanics had driven *that* idea into a ditch and thrown physics
into a quandary from which it has yet to emerge. When physicists
talk about their search for "a theory of everything," they are talking
about a theory that will reconcile the widely accepted but seemingly
contradictory ideas of Einstein and Planck. We don't have one yet.

Science is provisional, standing only until it is successfully challenged. Over the ages, quite a lot of scientific wisdom has been
proved wrong. The Earth-centered universe of Ptolemy gave way
to the heliocentrism of Copernicus; the idea that illness is a matter
of "humours" out of alignment has given way to the germ theory
of disease, and a lot of geology bit the dust with the development of
the theory of plate tectonics.

The history of science is filled with ideas big and small that were flat-out wrong—witness the astronomer Edwin Hubble discovering the universe is not stable, or Barry J. Marshall and J. Robin Warren, the Australian physicians, demonstrating that most stomach ulcers are caused not by stressful living and spicy food but rather by a bacterium, *Helicobacter pylori*.

John Maddox, who edited the journal *Nature* for a total of twenty-two years, until 1995, was famously asked once how much of what appears in the journal—one of the most prestigious in the world—would turn out to be wrong. He paused to consider before replying, "all of it."

So, when I talk about science, this is the kind of thing I am talking about: looking in the natural world for answers to questions about it, testing those answers with experiment and observation, and accepting findings as tentative, standing only until they are successfully challenged.

How does this process actually work? Typically, a researcher makes a hypothesis and designs an experiment, or a series of observations, to test it. The results either support the hypothesis or undermine it. If the hypothesis holds up, others attempt to replicate the work, looking for flaws in the idea. Once a hypothesis has garnered enough support, it becomes widely accepted. Often, though, experiments undermine the hypothesis. In that case, the researcher modifies it—or starts all over.

This is what we call the scientific method. It is not a smooth process. It is full of blind alleys, dud ideas, counterintuitive results, and contradictions. When these occur, they do not suggest the process of science has failed. They show that it is working as it should.

For example, on its website Understanding Science, the University of California at Berkeley cites what happened when scientists reported that measurements from Earth-orbiting satellites seemed to show that, global warming theories to the contrary, the planet was cooling, not warming.[4] This was a puzzle until other scientists realized that satellites gradually lose altitude as they orbit. Taking altitude changes into account, "the satellite measurements were much more consistent with the warming trend observed at the surface."

There are those who say science is the only source of truth, a view that reached its apotheosis with logical positivism, the idea that we can know something is true only if it is logically or empirically demonstrated. This idea is, in a sense, what Popper was talking about when he said ideas must be testable, or falsifiable. He called his view *critical rationalism,* and in one way or another it prevails in research today.

But some questions are outside the realm of science. Researchers do not necessarily have the answer, or the way to an answer, for everything. In his book *Rocks of Ages* Stephen Jay Gould, the author and evolutionary biologist, wrote about science and religion. They are two different realms, or "non-overlapping magisteria," he said.[5] This idea, called *NOMA,* has its critics—some say that scientists should not separate their religious views from their research work—but I think it is a good and useful principle.

I encountered a real-life example of NOMA in the person of one of my neighbors on Chappaquiddick, Joe Murray. Dr. Murray, who died in 2012, was a plastic surgeon and pioneer of organ transplants. He led the team that performed the first successful human organ transplant in 1954. The donor was a young Boston man who gave a kidney to his identical twin brother, who was dying of kidney failure.

Murray's research in his lab told him the work could be done, he recalled.[6] But should it? Plenty of people said no, even the medical officers at Dr. Murray's hospital, who pointed out that the transplant would violate standard medical ethics in that it would subject the healthy brother to a major medical procedure with no benefit, and potentially major harm, to himself. Others thought it smacked of playing God. These are not issues Dr. Murray or anyone else could elucidate in the lab. So he took this question to Catholic, Protestant, and Jewish leaders in the Boston area; they told him to go ahead. In 1990, Murray won the Nobel Prize for his transplant work.

In a much smaller way, I confronted this issue a few years ago when I gave a talk at New York University about religion and science. The audience was full of people who wanted to discuss stem cell research, especially research on human embryonic stem cells, obtained through the destruction of human embryos (usually donations from fertility labs). At the time there was a hot debate going on about whether that research should go forward and whether taxpayers should finance it. I told my audience those were questions science cannot answer, because as far as I can tell, your answers depend entirely on your view of the moral status of the human embryo.

My audience did not like that reply. As is often the case, they wanted their idea—that the research should go forward—to be anointed as the "scientific" view. But science alone cannot decide that issue.

What happens when science is wrong? How does the disparate, inchoate institution of science change its collective mind?

Thomas S. Kuhn, an American physicist and philosopher of science, attempted to answer this question in his 1962 book, *The Structure of Scientific Revolutions.* The book, a landmark in the literature of science, describes the way we (and researchers) see the world through the prisms of paradigms, "some implicit body of intertwined theoretical and methodological belief that permits selection, evaluation and criticism."[7] (Or as Einstein put it, we can only see what theory allows us to see.)

The idea that evil spirits make people sick is a paradigm, a world view through which traditional healers can devise ways of treating their patients. The idea that germs make people sick is another paradigm. The idea that the Earth is the center of the universe, and the idea that God created everything we know in six days are also paradigms, discredited now with the advents of new paradigms, the Copernican solar system and the theories of evolution and natural selection.

For every paradigm, some facts are crucial. The paradigm may predict still other facts, and when experiments or observations confirm them, the paradigm itself is supported and confirmed.

Sometimes, though, inconvenient facts refuse to fit the paradigm. Then it begins to fail. For a while, the paradigm can be adjusted to accommodate inconvenient facts, but if there are too many inconsistencies, they will eventually call the paradigm into question and hinder its application. When even repeated tweaking of the paradigm leaves uncomfortable facts that refuse to conform, researchers start looking for alternate paradigms. They look for new data, or look at existing data in different ways, something Kuhn called "picking up the wrong end of the stick." In other words, they frame the situation in new ways.

When there are many anomalies, even its adherents must question the fundamental usefulness of the paradigm. This usually happens when anomalies inhibit application of the paradigm in important ways, transforming them from annoyances into big deals.

By now, hints of a new paradigm or paradigms may be emerging. If observations or experiments show the prospective replacement paradigm is faulty, it too must be abandoned or modified.

Of course, people are unwilling to abandon paradigms that have served them well and that fit well with their core beliefs. That is why people were so unwilling to abandon the idea that the Earth was the center of the universe and accept Copernican principles. It is why many people are unwilling to acknowledge that our carbon-based economy needs substantial alteration if we are to avoid dangerous changes to Earth's climate.

How do scientists test their ideas? First, they have to define them precisely. They generate a hypothesis. Then, they design a way to test their hypothesis, whether in the lab, in the field, or in the case of fields like string theory, through elaborate mathematical complications. They gather the data their tests produce and analyze them. Finally, they report their results to their colleagues—in lab conversations, departmental seminars, conferences, or the like—and then submit them for publication in a scholarly journal and wait for someone to shoot them down. They hope, of course, that other researchers will replicate their findings.

Sometimes, of course, they cannot. It is important for researchers to know when replication efforts fail—but reports of failed experiments do not always draw wide attention. In February 2016, the biotech company Amgen and Bruce Alberts, a biochemist at the University of California, San Francisco and a former president of the National Academy of Sciences, announced the formation of an online forum on which researchers can share reports of failed efforts to replicate the research findings of others.

The site will be housed on an open-access research portal, F1000 on a channel called Preclinical Reproducibility and Robustness.[8] "Amgen is seeding the publication with reports on its own futile attempts to replicate three studies in diabetes and neurodegenerative diseases and hopes other companies will follow suit," the journal *Science* reported.[9] For researchers, the value of a finding—especially a stunning finding—emerges only as other researchers replicate it in a process that is a hallmark of science.

How Science Knows What It Knows

One of my colleagues in the Science Department of the *Times* reacts the same way whenever she hears someone making a startling scientific or engineering claim. She says, "Show me the data." My own response is similar. I say, "How do you know?"

The answers lie in the typical methods of scientific investigation. These methods can be divided roughly into two types: observation and intervention. That is, researchers devise a hypothesis and by observation or manipulation determine whether the facts on the ground (or under the microscope or through the telescope) are in accord with what their hypothesis would suggest.

In an *observational study,* researchers focus on their research subject, be it a group of people, an ecosystem, a mechanical device, or whatever, and observe what happens as it lives, changes, or operates.

The simplest observational studies are called *case series studies.* They are little more than a list of examples of the phenomenon of interest, whether it is fish exposed to a particular substance or people with a particular disease. Case reports can identify important phenomena. For example, the disease we now know as AIDS first emerged in the American medical literature in 1981, with reports of five young men in Los Angeles with an unusual form of pneumonia.[10] But case series studies cannot demonstrate cause and effect. They can only raise questions about whether such associations exist.

Case control studies differ in that they also include a control group without the particular characteristic of interest. For example, a case study of lung cancer might include a group of people with the condition and a control group of otherwise similar people without it. Researchers conducting such a study would attempt to see how the people with the disease differed from the people without it, in hopes that some interesting associations—like cigarette smoking—would pop out of the data. Then, they can test whether the association is meaningful—whether cigarette smoking actually causes lung cancer—or whether the association is a coincidence or the result of some other factor. This approach was what led researchers to link smoking to lung cancer. Although people often do it, it is also unwise to draw conclusions from case control studies. Like case series, their real use is generating hypotheses; they raise questions to answer in subsequent research.

Then there are *cohort studies,* in which a researcher starts with a group of research subjects with a given condition or exposure to a given substance or other characteristic, and then follows them over time to see whether they have a predicted outcome. Again, this kind of observational study—a study that says, for example, that factor A seems to be associated with factor B—does not prove that the two are locked in a cause and effect relationship. It is merely a clue to researchers that there may be something worth investigating.

For example, researchers who studied the decline in birth rates in Britain and Germany during and after World War II noted that a decline in stork populations occurred at the same time. Did a falloff in storks cause a falloff in babies? Probably not. Or take Seventh Day Adventists who, on average, live longer than members of other religious groups. Is that a sign of divine favor? Maybe. Or maybe it results from the fact that observant Seventh Day Adventists do not smoke. In other words, their longevity may be just another way of saying "smoking is bad for you."

To demonstrate cause and effect, as the Oxford statistician (and smoking-cancer theorist) Richard Peto famously put it, there is simply no serious scientific alternative to the generation of large-scale randomized evidence.

This kind of evidence is gathered in *interventional studies* or trials, in which researchers divide their subjects into two or more groups, randomly subject them to the various interventions of interest, like another treatment, a dummy pill or other placebo or, if it is ethical, no treatment at all (that's the control part). In medicine in particular, neither the participants nor the investigators providing the treatment know whether any particular subject is receiving the thing being investigated or something else—in that case the study is described as double-blind.

Designing this kind of trial can be an ethical and logistical nightmare, especially when the research subjects are human. For example, if scientists want to test a new drug, they must have enough faith that it works to justify giving it to some experimental subjects, and enough doubt about whether it works to justify not giving it to others. This balance is called "equipoise," and determining whether it exists is one of the chores of institutional review boards, or IRBs, which must put an ethical stamp of approval on research involving people.

In all this, the researchers are testing what is called "the null hypothesis"—the idea that whatever effect or association they are hoping to demonstrate does not exist. They look at the data they collect and calculate the odds that they would turn up that way, given the null hypothesis. Experiment or observation makes findings by disproving the null hypothesis.

It is a convention of research that a result that has only a 5 percent (or less) chance of occurring by chance represents something going on in the real world—that the result is, as researchers say, "statistically significant." This figure is called the "p-value"—if the p-value is less than 0.05, chance is considered an unlikely explanation for the findings.

The p-value reflects the strength of the association and the sample size. Even a small difference will achieve statistical significance if the study scope is large enough. Even a seemingly large effect will not be statistically significant if the study population is small. The smaller an effect, the harder it can be to demonstrate.

It is important to remember, though, that statistical significance does not eliminate the possibility that the results observed *did* occur by chance. And though the 95-percent standard, a p-value of 0.05, is almost universally used in research, it is arbitrary.

Many researchers now prefer to speak in terms of "confidence intervals"—the range of values within which they are 95 percent confident the true value lies. The width of the confidence interval reflects the precision of finding.

Also, it is important not to make too much of small-scale results. When a finding is significant but tiny, it can be "about as meaningful as a coin being flipped fifty times and not coming up heads half the time," John Allen Paulos, a mathematician at Temple University, wrote in his book *Innumeracy*. "Too much research in the social sciences, in fact, is a mindless collection of such meaningless data."[11]

There are a few ways researchers can—deliberately or inadvertently—design a study to produce a desired result. A study can be kept so small that any effect it demonstrates will lack statistical significance—a practice known as obtaining a reliably negative result.

A study of something with a long latency period—cancer, say—can be conducted over such a short time that the disease would

not have time to develop. Or people studying the effects of exposure to something might mix subjects with high and low exposure,
diluting the measurable effect.

It is also possible to interpret data to maximize the chances of its
passing the test of statistical significance. John Ioannidis, a professor
of health policy at Stanford, calls this practice "significance chasing."
He summed up his conclusions in a 2005 paper, "Why Most Published Research Findings Are False." The paper, published in *PLoS
Medicine,* an online journal, has become a kind of cult classic.[12] The
essay asserted that many—in fact even most—published research
reports are marred by miscalculation, poor design, or misguided
analysis of data. "There is an increasing concern that in modern
research, false findings may be the majority or even the vast majority of published research claims," Ioannidis said.[13]

He is not the first to make this kind of charge. For example, a
few years earlier researchers from the University of Girona in Spain
reported that their study of a sample of papers from the journal
Nature found that 38 percent contained one or more statistical errors. A quarter of papers in the *British Medical Journal* had the same
problem.[14]

Later, *PLoS* published two more essays in reaction to Dr. Ioannidis's argument. Neither of them refuted it. The first noted merely
that the probability that published research will turn out to be
wrong diminishes as other people replicate it. The second made the
point that the more people want the information, the more they
may accept "less than perfect results."[15] One remedy might be for
researchers to publish their raw data as well as their results. Now
that this kind of information can be posted relatively easily online,
the practice is becoming more common.

Researchers also have to think hard about how their results
might be altered by factors outside the scope of their experiment or
observation—so-called confounding factors. For example, mortality rates are lower in Alaska than they are in Florida. Does this
mean Alaska is a healthier place to live? Or does it mean there are
more old people in Florida?

In a less obvious example, when researchers look at the value of cancer screening, they usually find differences in disease outcomes between people who undergo regular cancer screening and those who do not. Usually, studies show that those who are screened do better. But these findings must be viewed cautiously, because people who on their own undergo regular health screenings tend to be "richer, insured, with access to the health care system, health-conscious, healthier generally," as Dr. Barnett S. Kramer, an expert on medical evidence, noted at the 2003 conference on the subject at the Massachusetts Institute of Technology. They are "health-seekers," he said, and comparing them to people who don't or aren't able to take such good care of themselves may produce invalid results.

At the conference, Kramer told us that in clinical trials people who follow the prescribed regimen often do better than those who do not, *regardless of whether they are taking an actual drug or a placebo*. He said other data show that women who sign up for regular screening for ovarian cancer die less often from colorectal, stomach, lung, cervical, and other cancers—not because the ovarian cancer screening is useful against those cancers or even for ovarian cancer (we still do not have a good screening test for ovarian cancer), but because women "who chose to show up for screening are fundamentally different from the ones who didn't."

If the study involves food, maintaining true placebo or control groups can be impossible, unless trial participants are sequestered in sealed rooms, a step rarely taken because of its hideous expense and inconvenience.

Sometimes results are skewed by selection bias, which occurs when study participants are not chosen randomly. For example, a study whose subjects were people who responded to an online survey is obviously hampered by selection bias—its subjects are said to be "self-selected" rather than identified at random.

A more important issue relates to the fact that positive results—findings that disprove the null hypothesis—are more likely to see the light of day in a scholarly journal than findings that force researchers to say, "oh well, better luck next time." At the *Providence*

Journal, the Rhode Island newspaper where I began my journalism career, we used to call this kind of thing a "no kangaroo in the Biltmore lobby" situation. If you discover a kangaroo in the lobby of the hotel, you have a story. No kangaroo, no story.

Researchers call this problem publication bias. Like newspapers, publishers of research journals prefer positive findings, probably because their readers find them more interesting and important. Negative findings tend to be filed away in obscurity, which is why researchers also call publication bias "the file drawer problem." The relative absence of negative findings from the scholarly literature is important in all fields, but it is particularly troubling when it comes to clinical trials—the knowledge that a trial showed a drug or other treatment had no effect should be in medicine's knowledge bank somewhere.

Nancy L. Jones, a policy analyst at the National Institute of Allergy and Infectious Diseases, neatly summarized these issues in an essay on training young researchers in the ethics of their profession. She published the essay in *American Scientist,* a publication that never fails to fascinate, produced by the scientific honor society Sigma Xi.[16] Scientific progress hangs not just on the publication of new findings, she wrote, but on reports of "negative results and repetitions of previous studies." Reporting negative findings can steer other researchers away from traveling down blind alleys. (Some argue, Jones wrote, that minimizing these experimental journeys is a moral imperative, especially when the potential experimental subjects are people or other primates.) And because it is not the initial finding that persuades researchers, but the meticulous, accumulated repetitions of the same result, "there must be a place to report follow-up studies that confirm or refute previous findings."

Instead, Jones said, the growing difficulty researchers face in obtaining funding has "placed the scientific community in an advocacy role. They are no longer just the guardians of knowledge; they compete for public resources and champion their specific fields." Jones said they may find themselves "promoting the potential outcomes of research—such as cures, solutions and new economic streams—rather than justifying support for the research itself. The scientific record is not immune to this pressure. Scientific societies that publish journals can be tempted to boost the prestige of their fields by prioritizing highly speculative, sexy articles and by egregiously over-promoting the potential impact of the research."

The over-promotion of research became such an issue for the journal *Nature,* a for-profit enterprise, that it began routinely issuing "hype" alerts in the press releases it sends to journalists about the contents of the coming issue, asking reporters to alert the editors if they felt a finding had been overhyped.

Researchers are also more likely to report results that support their theories, especially if they have a financial stake in the outcome, or they expect promotion, tenure, or other professional advancement as a result of publication.

Also, research suggests that peer reviewers, the experts who review articles before their publication in scholarly journals, are more likely to give a thumbs-up to a paper that reports a positive result. According to *Nature,* "Unless a paper convincingly overthrows a widely held belief, negative findings tend to be of less interest than positive ones."[17]

That may be putting it mildly.

Seth Leopold of the University of Washington Medical Center in Seattle reported similar findings at a 2009 meeting of the International Congress on Peer Review and Biomedical Publication. He described what happened when he and his colleagues asked reviewers at two orthopedic journals to rate a bogus manuscript. In one version, the manuscript reported that one antibiotic regimen did a better job of preventing infection than another one. The second version reported that the treatments were equally effective. Almost all the reviewers asked to rate the first paper deemed it worthy of publication; only 71 percent of those reviewing the second paper felt the same way. Leopold also reported that reviewers were more likely to praise research methods supposedly employed by the researchers who wrote the first bogus paper, and were less likely to spot intentionally introduced errors.[18]

Some researchers think publication bias may lie behind European reluctance to embrace genetically modified crops. Trials of the crops "overwhelmingly reveal no adverse environmental consequences," *Nature* argues, but these results generally go unpublished.[19]

Publication bias is a serious problem. For one thing, people may spend a lot of time unknowingly repeating work that has already shown negative results. Worse, *Nature* wondered, "is our scientific understanding in some cases biased by a literature that might be inherently more likely to publish a single erroneous positive finding than dozens of failed attempts to achieve the same results?"

As you can imagine, this issue is a particular problem when researchers sweep the published literature for findings on a particular question and merge disparate studies in an analytical technique called meta-analysis. If negative results are not among the published papers, it is possible that results aggregated from published papers constitute a biased sample.

In the 1990s a number of studies were reported showing a link between cervical cancer and use of oral contraceptives. But an analysis in 2000 showed that studies finding no link—"negative results"—were seldom published. Result: anyone searching the scholarly literature for information on any link between cervical cancer and the use of oral contraceptives would come away convinced that the weight of evidence supported a connection. But it did not.[20]

Some researchers call for the creation of registries where scientists could see results from all clinical trials. Jonathan Schooler, a psychologist at the University of California, Santa Barbara, made a similar plea in an essay in *Nature* in 2011. "Publication bias and selective reporting of data are similarly difficult to investigate without knowing about unpublished data," he wrote.[21]

The problem eventually provoked researchers to establish journals dedicated to the publication of negative results, such as the *Journal of Negative Results in Biomedicine* and the *Journal of Negative Observations in Genetic Oncology*. The latter journal is particularly useful because while findings of connections between cancer and one gene or another turn up regularly, findings of no link do not.

Sometimes it is relatively easy to answer a research question. For example, if the question is whether a given microbe causes a given infectious disease, effects of exposure are usually seen quickly. There is a known mechanism by which the microbe does its dirty work, and even a set of rules, Koch's postulates, for establishing a cause-and-effect relationship between a given microbe and illness.

But with conditions that take a long time to develop, it's not so simple. Some of these conditions are diseases, like cancer or heart disease; some result from exposure to environmental pollution. Smoking or eating fatty food or dumping waste in the river may continue for years or even decades before it will cause an observable problem. Conditions with such long latency periods are difficult—and expensive—to study.

In that event, researchers may fall back on what are called surrogate end points. With cancer, for example, what we care about most is whether people will die of the disease. Because cancer usually has a long latency period, we use detection of cancer as a surrogate end point. But as scientists have learned, merely detecting a cancer does not tell you whether it is potentially lethal, or indolent and more or less harmless.

This effect appears vividly with prostate cancer. In recent years, we have become much better at testing for it and much more vigilant, but the prostate cancer death rate has barely budged. The situation is somewhat similar for breast cancer—its death rate is down a bit since screening mammography became widespread in the 1980s, but that decrease must be attributed also to big improvements in chemotherapy and other treatments.

With pollution issues, to cite another example, what we want to know is whether the pollution is causing harm. The mere fact that a pollutant can be observed in the environment does not mean it will inevitably cause a problem. The presence of the substance is a surrogate end point.

In a way, using surrogate end points is like measuring the effectiveness of a job training program by the number of people who signed up for it, when what you really want to know is how many participants got a job and held it.

The moral: ask yourself whether the thing being measured or detected is the thing you really care about.

Perhaps the most troubling problem category belongs to ideas we do not even think to question because we know they must be correct.

For example, enforced sterilization of the supposedly "unfit" was supported by the ideas of eugenics, mainstream at one time but discredited today. (Creationists often cite eugenics as evidence that science is an imperfect enterprise—which of course it is.)

Not too long ago, the medical establishment listed homosexuality as a mental illness; today gay marriage is the law of the land.

At the *Times* we encountered flat-out-wrongness as we covered the anthrax attacks that occurred after 9/11. As the attacks were underway, doctors from the Centers for Disease Control and Prevention told us what to expect from the disease. The news was bad: if you inhale anthrax spores you might not know you have contracted inhalation anthrax until you start showing symptoms. And by then it is too late to save you, because the bacterium's toxin is already at work in your body. But there was (relatively) good news too: you would have to inhale 10,000 anthrax spores at least to have a 50 percent chance of contracting a lethal dose. (As epidemiologists would say, the LD_{50} of inhalation anthrax is 10,000 spores.)

News organizations around the country, including the *Times*, reported these assertions with confidence. But they were wrong. As it turns out, antibiotics can save people with inhalation anthrax, even after symptoms emerge. And as the deaths of a Connecticut pensioner and a Bronx hospital worker showed, people can contract a lethal dose from very few spores. How did the government's experts get things so wrong?

Anthrax, particularly inhalation anthrax, is extremely rare in the United States, so the CDC doctors were basing their conclusions on old data, from an outbreak in a New Hampshire woolen mill in 1957 and the accidental release of anthrax from a Soviet bioweapons lab in 1979. In neither case were conditions remotely like those encountered in the fall of 2001. Also, there had been decades of improvement in drugs and other treatments. But the CDC, like everyone else, was learning as it went along.

The *Times* confronted a similar problem in the run-up to the 2003 invasion of Iraq, when the Bush administration was making a big deal out of the possibility that someone could attack the United States with smallpox. Routine vaccination for the disease ended in the United States in the 1970s, and the disease was declared eradicated in 1980. Nevertheless, the Bush White House said, someone like Saddam Hussein might have a secret stash of virus. "Smallpox martyrs" might allow themselves to become infected and then travel to and through the United States, spreading the disease as they went, until they succumbed to it. Citing government statistics, we reported that smallpox has a mortality rate of about 30 percent. That figure turns out to be wrong on two counts.

In the years since the disease was eradicated, there has been a lot of progress on antiviral drugs, some of which was motivated by the advent of AIDS. So doctors have much more to offer smallpox victims than they once did. Yet there are many more people walking around today who would be extraordinarily vulnerable to the disease, because their immune systems have been damaged by age, cancer treatment, immunosuppressive drugs they take after organ transplants, or by AIDS. Fortunately, a smallpox attack did not materialize. And after the United States invaded Iraq, the Bush administration stopped talking about smallpox.

Models

Orrin H. Pilkey Jr., a geologist at Duke University, has an idea for the coastal engineers who design beach nourishment projects, in which dredges pump sand onshore to restore eroded beaches. He says that instead of drawing up elaborate plans for the work, with mathematical models based on calculations of the natural slope of the beach, the size of its sand grains, its prevailing winds and currents, and so on, engineers should simply instruct workers to dig up a lot of sand somewhere and dump it on the beach.

Pilkey, a longtime critic of many beach renourishment projects, calls this approach "kamikaze engineering."[22] He acknowledges it might not do very well. But, he often points out, beach nourishment projects engineered to the hilt don't always do very well either. In spite of the elaborate models predicting how long the additions of sand will survive, and therefore whether they are worth what they cost, projects in the United States have been plagued by problems. One notorious project in Ocean City, Maryland, washed out in a storm before the equipment used to build it had been removed from the beach.

But by the time storms or erosion wipe out one of these renourishment projects, people are no longer thinking about the model that assured them it would last a long time. Instead, project backers describe storms and erosion—inevitable on the coast—as "acts of God" no one could have predicted. Or they describe the projects as "sacrificial": though they fail, they have protected shoreline buildings. And efforts begin to summon up tens of millions of state, local, and federal dollars for the next project to protect the hotels, condos, and beach mansions built too close to the tidemark for their own good.

The advantage of the kamikaze approach, Pilkey says, is that it would not lull anyone into a false sense of engineering security. It would make clear what anyone who pays attention to the issue knows: constructing a mathematical model of a beach is a fiendishly difficult enterprise, and our record of success at it is so slim we might as well acknowledge we cannot yet do it.

Mathematical models are widespread in science and engineering research. Sometimes they work well. Sometimes they do not. Models typically process data about measurable variables—like wind direction, currents, and so on—through mathematical formulas to arrive at what designers call "quantitative outputs." Typically, these outputs tell us what is likely to occur in the future if this, that, or the other thing happens today.

Whether the issue is climate change, pollution abatement, the possible spread of bird flu, or the effects of tax cuts on economic growth, researchers have a model for it. The weather forecast is, in effect, the fruit of models. Advocates of all stripes rely on models as they push for policy action.

Often, there are no good alternatives to models. The Office of the Science Advisor to the Council for Regulatory Environmental Modeling of the Environmental Protection Agency stated in guidance it issued for modelers in 2009, "the spatial and temporal scales linking environmental controls and environmental quality generally do not allow for an observational approach." That is (I think) another way of saying it would take way too long and cost way too much money to do experiments in the field. Therefore, we need models.

"When we are working in the future, there is no possibility of frequency statistics," the climate expert Stephen Schneider, who died in 2010, once put it in a presentation at a meeting of the American Association for the Advancement of Science.[23] "In the future, there is no data," he said. "All there is, is models." As he noted, the alternative to a model is ordinarily the assumption that whatever trends have prevailed in the past will continue into the future—itself a kind of model. So it is likely that we will continue to be urged to adopt policy choices or make other decisions based on data from models. When you want to evaluate such advice, you will have to evaluate the models.

The most important thing to know about models is this: they are simulations produced by solving equations based on a set of assumptions. On their own, they can demonstrate at most the implications of the assumptions on which they are based. They do not necessarily show that a phenomenon you care about is created in real life in the manner the model suggests.

In their book, *Useless Arithmetic,* Pilkey and his daughter Linda Pilkey-Jarvis argue, in effect, that nature is too complex to be usefully modeled in any detailed way. As a result, they argue, policy-makers should use models to discern qualitative trends rather than to predict precise quantitative results. They say it is virtually impossible to produce quantitative models that will accurately predict the outcome of natural processes on Earth's surface.[24] Modeling to predict the performance of renourished beaches is their prime example, but there are plenty of others.

One is the modeling of the movement of nuclear waste through an underground storage site at Yucca Mountain in Nevada. The Pilkeys cite it not because the site has failed—it has yet to be built and its future is in doubt—but because of the unrealistic expectations people built into the modeling process.

At first, modelers were required to predict what would happen to the waste over 10,000 years, far longer than recorded human history; later, there were demands for million-year safety guarantees. That is beyond the capacity of any model. In fact, the requirement is so unmeetable it can be viewed as little more than a way of killing the project.

The much-modeled but ultimately collapsed Northwest Atlantic fishery for cod and other ground fish is another modeling failure. For years scientists made regulatory recommendations based on mathematical models of cod stocks. Today, the cod fishery in maritime Canada is closed and in New England, home of Cape Cod, scientists say it teeters on the brink. (Fishers don't necessarily agree; they don't trust the scientists' models!)

The most controversial models are the ones climate experts use to predict the effects of growing quantities of greenhouse gases in the atmosphere. The world's scientists, virtually to a person, are confident that the models are correct to predict, overall, that unless people change their use of fossil fuels the planet is heading for disaster—driving at full speed toward a brick wall, as Prince Charles famously put it.

There are five major climate models in use today.[25] Basically they work by dividing Earth and its atmosphere into a global three-dimensional grid (the size of component compartments has shrunk over the years as computer power has increased). Then the researchers set initial conditions in each part of the grid and then set climate-related processes to work. Among them are the movement of surface and groundwater, wind, the presence and influence of pollutants in the air, the influence of clouds and water vapor, carbon cycling (the growth of plants, which take up carbon dioxide, but whose growth is influenced by factors like rain), evaporation of moisture from the land and from the oceans, ocean circulation, the melting of glaciers and inland ice sheets, and even the melting of Arctic sea ice.

All of this is highly complex. For example, clouds and water vapor trap heat, but they also reflect solar radiation back into the atmosphere, with a cooling effect. Clouds and water vapor are believed to be central to climate change—but most people agree our models for them are inadequate.

Overall, though, climate models agree that things are getting warmer. That is not a surprise. There is no doubt that the chemistry of the atmosphere has a lot to say about the climate on Earth, and there is no doubt that people have been changing that chemistry in a big way since the beginning of the Industrial Revolution 150 years ago, chiefly by increasing atmospheric concentrations of heat-trapping gases like CO_2 and methane. The idea that we could do that without consequences to Earth is farfetched.

Climate deniers seize on differences in model predictions and legitimate doubt about particular aspects of particular models to raise the flag of uncertainty about the climate issue as a whole. This tactic is disingenuous, but legitimate climate researchers have had trouble countering it. In fact, they are quick to admit that their models are highly imperfect. Given that, is it reasonable to rely on the mathematical calculations and computer simulations upon which the world's climate experts base their forecasts? I think it is, in large part because the climate experts understand the limits of their models and do not press them to say more than they can say. For example, "climate models do a poor job when it comes to simulating sea-ice changes in the Arctic." That was the conclusion of researchers at the Geophysical Fluid Dynamics Laboratory in

Princeton who compared data from satellite observations with results from five models described as "state of the art." In all cases, the researchers found, the models "considerably underestimated the sea-ice decline."[26]

That, incidentally, has been a pattern. Even as climate deniers pooh-pooh the predictions of climate scientists, events in the real world suggest their predictions are, if anything, optimistic. In what many researchers called an excess of caution, in 2007 the Intergovernmental Panel on Climate Change rejected model findings on the melting of inland ice sheets and glaciers. As a result, they reduced their projections for future sea level rise. The panel's previous estimate was one meter—about three feet—by 2100. The 2007 report cut that back to sixty-six centimeters—about two feet. The reason was not any kind of doubt on the experts' part about the reality of human-induced climate change and the way it fuels rising sea levels. If anything, they believed the rise was accelerating. The problem, they said, was their models.

Sea level rise is driven by the expansion of water as it warms (a well-understood and relatively easy-to-model phenomenon) and the melting of inland ice sheets and glaciers, which sends water into the sea. This melting ice is by far the greater potential menace, but modeling it is difficult. So in their 2007 assessment the IPCC basically left it out. This move produced a bizarre situation in which the report scaled back estimates of sea level rise, even as the researchers who wrote it believed the rise was accelerating.[27]

The climate scientists, like responsible modelers generally, are up front about the flaws of their models. As Schneider said at the AAAS meeting, "what you have to do is be as transparent as you can."

The modeling process is further complicated by the tremendous growth in data generated by ever more sensitive sensors and other devices. The computer software necessary to marshal this data is often highly specialized, and often written ad hoc by the researchers doing the modeling work. Critics of over-reliance on modeling argue that this process can produce overoptimistic assumptions and oversimplified data that lead to erroneous results.

Some say this kind of thing even contributed to the troubles Wall Street's "quants" experienced in the financial collapse of 2008. "There was willful designing of the systems to measure the risks in a certain way that would not necessarily pick up all the right risks," said Gregg Berman of RiskMetrics.[28] Among other things, some models were designed to assume the market placidity of the previous few years was normal and would persist, or described arcane new financial instruments as if they were standard bonds with known interest rates and maturity periods—and track records. On Wall Street, newly created financial instruments are usually regarded, almost by definition, as more risky.

"Researchers are spending more and more time writing computer software to model biological structures, simulate the early evolution of the Universe and analyze past climate data, among other topics," the journal *Nature* noted in October 2010. "But programming experts have little faith that most scientists are up to the task." The article cited James Hack, director of the U.S. National Center for Computational Sciences at Oak Ridge National Laboratory, who said, "The level of effort and skills needed to keep up aren't in the wheelhouse of the average scientist."[29]

Also, it is rare for this kind of computer code to be made public. Although research papers may include "a brief mathematical description of the processing algorithm, it is rare for science software to be published or even reliably preserved," Nick Barnes, director of the Climate Code Foundation, wrote in the same issue of the journal. "The open-source movement has led to rapid improvement within the software industry," he added. "But science source code, not exposed to scrutiny, cannot benefit in this way."[30]

In other words, if researchers published the computer code they used to analyze their data, programmers could point out flaws that might compromise the reliability of the researchers' results.

But some researchers are reluctant. In many disciplines sharing code is just not done. Some researchers fear that if they publish their code, they may be criticized for its flaws or even hounded for technical support by others who want to use it. And, of course, some people may be reluctant to simply give away a tool it took a lot of time and effort to develop.

To these reluctant people Barnes poses a question: "Does your software perform the algorithm you describe in your paper?" If it does, he writes, your fellow researchers will accept it. If it doesn't, "you should fix it anyway."

How big are problems with any particular model? A lay person may find that question impossible to answer. But it is something to think about.

Modelers must be sure, as Schneider put it, that their models are based on data that "represent the real world, and not just the data collection process."[31] But in many cases scientists or engineers face intractable difficulties getting good data. For some models, climate and beach erosion, to name two, too much data is simply missing

Finally, in the real world, decisions often must be made before data or models are complete. Nanomaterials are already making their way into the environment in a host of consumer products, but our knowledge of whether and how they will enter the water supply can be based at this point only to a very limited extent on real world data. The rest comes from models. The best we can hope for here is that unknown unknowns will be converted into known unknowns; in other words, that we will at least know what is missing. But in nature there are always unknown unknowns.

Given the inevitable weaknesses of most models, responsible researchers anticipate that new information may challenge the premises on which their models are based. They design their models to be adaptable to new information. In the environmental realm, revising assessments (and therefore, in theory, policy) as new information comes in is called adaptive management, and it is an approach people like Pilkey endorse. For him, it is the only context in which environmental action on model recommendations is acceptable.

Schneider believed that when scientists present model data to policy-makers—and by extension to the public—they should insist on requirements that the models be revisited at least every five years. "Politicians don't like that," he said, "but it's adaptive management."

Unfortunately, modifying or even abandoning plans in the face of new data can be expensive. Also, it requires assiduous monitoring. Often, initiatives buttressed by modeling data are undertaken with guarantees that the results will be monitored. Often the required monitoring does not take place, for a number of reasons.

For one thing, monitoring can disclose—sometimes quickly— that things are not working out the way people planned. Generally, this is information no one wants to receive, especially if they have to pay for it.

One form of monitoring is what the federal Food and Drug Administration calls postmarketing surveillance—the tracking of people who take a particular drug after it has been approved for sale to see if unexpected side effects turn up once the drug goes into wider use. This kind of monitoring can turn up information that can force drugmakers to pull a profitable product off the market.

When it comes to beaches, advocates of regular multimillion-dollar replenishment projects have little to gain by documenting just how quickly the projects fail.

Anyway, the excitement in science or engineering is discovering a previously unknown phenomenon or designing a new process or device, not necessarily watching what happens next. "Monitoring is science's Cinderella, unloved and poorly paid," Euan Nisbet, an earth scientist and atmospheric chemist at the University of London, wrote in the journal *Nature* in 2007.[32]

But it is immensely valuable. Consider the effort Charles David Keeling began in 1958 on the Mauna Loa volcano in Hawaii. He monitored levels of carbon dioxide in the atmosphere, and the resulting clear evidence of its accumulation has been a critical piece of evidence in climate research. Keeling's chart of his data, the Keeling Curve, is an icon of science. At least initially, however, the effort had little support. It almost died in 1964 for lack of funds.

Merely tracking something as it unfolds does not inspire funding agencies or anyone with commercial ambitions. As Nisbet puts it, "funding agencies are seduced either by 'pure' notions of basic science as hypothesis testing, or by the satanic mills of commercial reward.

Neither motive fosters 'dull' monitoring because meeting severe analytical demands is not seen as a worthwhile investment ... *In situ* work promises neither shiny rockets nor lucrative contracts. Monitoring does not win glittering prizes. Publication is difficult, infrequent and unread."[33]

I encountered this issue, in a way, when I wrote an article about scientists at Harvard and Boston Universities who were studying plant life in Concord, Massachusetts, the town famous for "the shot heard round the world" in the American Revolution but also the home of Walden Pond and Henry David Thoreau.

Thoreau kept meticulous records of plant life in town, especially noting when flowers appeared in the spring. Drawing on those records, the plant collections at the Harvard University Herbaria, and notebooks of other eccentric nineteenth-century Concord naturalists, the researchers concluded that spring arrives in Concord a week earlier now than it did in Thoreau's day.

Those researchers were lucky. The notebooks were gathered and preserved at the Concord town library—even though hardly anyone ever consulted them—and Harvard has space and staff to preserve a vast collection of plant samples. Nowadays, the researchers said, it is hard to defend spending money on this kind of thing.[34]

How can you assess the usefulness of a model? First, consider whether the model is quantitative or qualitative. Is it suggesting how a trend is moving (sea levels are rising), or is it offering precise mathematical predictions (a beach replenishment project built along certain lines will last so many years)?

Then consider the factors the model relies on, its parameters. Consider how sensitive the model as a whole is to changes in one or more of these parameters—that is, how likely is it that changes in the parameters change model results? Keep in mind, though, that assessing this kind of model sensitivity is a judgment about the importance of the parameter to the *model,* not necessarily the importance of the parameter to the real world. As the Pilkeys note, if a model "is a poor representative of reality, determining the sensitivity of an individual parameter in the model is a meaning-less pursuit."

Also, consider the availability of the data the model needs. Keep in mind that some things can be measured directly, accurately, and quickly and some cannot.

Often, if you dig a bit into model design, you will see factors seemingly floating in the calculations—often called "coefficients" of one thing or another. Critics of modeling say modelers can inject coefficients into their models in order to make the models produce desired results. They jokingly call this process "objective analysis"—tell me your objective and I will give you an analysis that supports it. Coefficients are not necessarily evil. But it is good advice to look for coefficients and satisfy yourself that their presence is legitimate.

Also remember that even the best model depends on the reliability of the data fed into it (and the reliability of the data that existed when the model itself was invented). If you can, find out where the data come from and consider whether the source is trustworthy.

In 2009 the office of the science advisor of the Council for Regulatory Environmental Modeling, an arm of the Environmental Protection Agency, issued "guidance" on the construction of models that would be useful for policy-makers.[35] The resulting document is almost incomprehensibly complex, at least to me. But though it was obviously not directed to a lay audience, it offered some information to keep in mind when you are trying to evaluate claims or proposals based on models.

What you really need to know can be summed up like so: Are the model's underlying scientific foundations sound? How reliable are the data? That is, how free are they from bias owing to limited sample size, analytical imprecision, and so on?

Here are some other things to consider:

- Who paid for the model?
- Was the model based on good scientific principles?

- Look at its assumptions. Is enough known about the situation to stamp each as reliable? If not, the model is weak. As Pilkey and his co-authors asserted in a 2000 research report, "each important model assumption must be examined in isolation; incorporating them into a model does not improve their validity." In an ideal world, each parameter (variable) in a model should be accompanied by estimates of the probability of its occurrence. This information may be difficult to obtain. But if you cannot obtain it, you know at least what you are missing.
- How well has the model been corroborated by comparison to real world data?
- What was the source of the data used to design the model?
- How closely does the model approximate whatever it is you actually want to know about? A model might do an excellent job approximating the effects of one variable or another, but fail to reflect the situation as a whole.
- Is the process of model development well documented?
- Has it been subjected to peer review? Peer review "is not a truth machine, not even close," as Schneider said, "but it does have some value."
- Where do the modelers come from? There are those who say a model developed entirely in one lab may be weakened by the absence of alternate perspectives. Modelers must be "open to other points of view," as Schneider put it.[36]
- Some models can be tested by assessing how well they predict what actually happened in the past, a process called hindcasting. In other words, feed some initial conditions into the model—the characteristics of a beach in 1950, say—start the model running, and see if its description of the beach today matches what it really looks like. If the model you are interested in has been subjected to hindcasting and has held up, that's a good sign.

- Finally, once a model is in use, you should look for indications that people are monitoring its performance and are ready to revise or even scrap it in light of new information.

In theory, the combination of increasing computer power and a growing supply of real-world data should lead to better and better models. Still, "Beware of big complicated models and the results they produce," Jonathan Koomey advises in *Turning Numbers into Knowledge.*[37] "Don't be too impressed by a model's complexity. Instead ask about the data and assumptions used to create scenarios. Focus on the coherence of the scenarios and their relevance to your decisions and ignore the marketing doublespeak of those whose obsession with tools outweighs their concern with useful results. Sadly, many modelers fall into this latter camp, and you would do well to avoid them."

A Jury of Peers

When I was science editor of the *Times,* Lawrence K. Altman, our chief medical writer, often told me he wanted to write about peer review, the checking process scholarly reports must undergo before they are published in reputable (that is, "peer-reviewed") journals. I always resisted, on the grounds that while the subject was worthy, it was dull. It was too much like what we journalists call "inside baseball," of probable interest only to the people who already know all about it.

But Larry was right. Few processes in science have more implications than the operation of peer review, the review of one's scientific work by one's scientific peers. Few are more mysterious and few are more open to manipulation.

The process begins when a research report is submitted to a journal for possible publication. If the editors of the journal think it looks interesting, they send it out to other experts for their assessment. In general, reviewers determine whether the researchers designed their project well, ensure they have not ignored important relevant ideas, and check that their conclusions make sense from the evidence they produce. Peer reviewers, typically volunteers, may suggest that manuscripts be accepted as is (an unusual outcome), accepted pending suggested revisions, or rejected out of hand.

In a way, peer review is the rock upon which the edifice of scientific and engineering research is built. Publication in a prestigious journal can make a researcher's career. At the same time, publishing papers that turn out to be influential—widely cited in subsequent research—is vital for the success of a scientific journal. Most scientists look askance at findings reported without the benefit of peer review. Everything from the conclusions of the Intergovernmental Panel on Climate Change to decisions in court can hang on whether a finding is peer-reviewed.

Peer review by anonymous outside experts did not take hold in science until the 1930s, and the system achieved something like its present form in the years after World War II. Advocates of peer review—and in the absence of a better approach I am one—would like to believe that ordinary people can trust the conclusions of peer-reviewed work. But even I recognize that is not necessarily the case. The peer review process is in trouble, drawing new and unwelcome attention. "This 'peer review' is supposed to spot mistakes and thus keep the whole process honest," the *Economist* noted recently. "The peers in question, though, are necessarily few in number, are busy with their own work, are expected to act unpaid—and are often the rivals of those whose work they are scrutinising."[38]

In part (but only in part) because they are unpaid, good peer reviewers can be hard to find, especially for papers with lots of data and supplementary material, which may be time-consuming, tedious, or otherwise difficult to assess. As a result, journals typically choose only two or three reviewers for each paper, and they may miss inaccuracies, especially if the work in question is multidisciplinary or interdisciplinary, in which case reviewers from multiple disciplines must be found. In some fields, like cutting-edge mathematics, the pool of people who understand the work is so small that it can be almost impossible to recruit reviewers.

The paucity of willing reviewers is such that William F. Perrin, a marine biologist at the Scripps Institution of Oceanography and a onetime editor or associate editor of scientific publications, suggested that in fairness researchers who publish three or four papers a year should agree to do at least twelve to sixteen reviews a year. "Anything less means that you re sloughing off the work to others who are perhaps less knowledgeable and capable than you in your specialty, and you should not be upset when someone reviewing a paper of yours 'doesn't have a clue.'"[39] But reviewing even a paper a month, if you take the chore seriously, is a serious time commitment.

Meanwhile, reviewers must take the writers' word for it that the research was conducted as described and that the data in the paper are genuine. After all, they were not standing at the laboratory bench or out in the field when the work was done. The entire process is built on trust and, unfortunately, sometimes that trust is misplaced.

That is what happened in 2005 when it was discovered that a South Korean researcher who had made a splashy report about cloning of a human embryo admitted that he faked the whole thing. The journal *Science*, which had published a report of the supposed work, retracted the paper. Later, Monica Bradford, then the journal's executive editor, assessed the episode. She called it "distressing," but said she did not think it reflected badly on the journal or how it had handled the review of the finding.[40]

Retractions are common in science, she said, and most result from innocent mistakes—"misinterpretation or human error that was accidental." She conceded, though, that "basically a reviewer has to begin with a position of trust, and everything they are seeing is real and not fabricated," she said. "And if someone is good at fabricating, it would be hard to know the difference."

Ordinarily, it is only when other researchers try to replicate the work "or build off it," she said, that problems emerge. "If that

cannot happen, scientists start to question the data and it is retracted, or people follow other lines of inquiry."

She said *Science* published between 800 and 900 research reports
per year and, on average, four or five retractions. "And most of them
would have been from human error." Peer review, she concluded,
"has proven to be the best system we have. For now, it seems to work."

But another scandal makes me wonder. It involved Dr. J. Hen-
drik Schön, a scientist at Bell Labs who from 1998 to 2001 published seventeen papers, with twenty co-authors, including claims
that he had created molecular-scale transistors.[41]

Plenty of people tried and failed to replicate Schön's work, but
his fakery came to light only when researchers noticed he had published identical graphs in different papers. (An internal investigation at Bell Labs cleared his collaborators, who in most cases had
merely contributed materials to the supposed work.) If he had not
fudged his diagrams, it is not clear how long it would have taken
for others to discover his misdeeds.

The choice of reviewers says a lot about whether a paper will be accepted for publication or rejected. Usually, reviewers are anonymous. In theory, that protects reviewers from the wrath of rejected
authors and enables them to offer freely their criticism or suggestions.
It also means they can recommend rejection with relative impunity
even if they have biases or grudges against authors, or regard them
as rivals.

The authors of research papers can avoid this outcome if, as is
common, they are asked to recommend reviewers and name anyone
they think would do a bad job. In principle, this is a good idea. Authors are typically well placed to know who is expert in their own
fields. On the other hand, they may have self-serving reasons to
keep someone—a rival, say—out of the process.

"Either suggesting or excluding reviewers, the studies show, can
significantly increase a manuscript's chances of being accepted,"
the journal *Science* reported, referring to research on the peer review process.[42] The journal cited work by Lowell A. Goldsmith, a

dermatological geneticist at the University of North Carolina, Chapel Hill, and editor of the *Journal of Investigative Dermatology*. He and his colleagues looked at 228 consecutive manuscripts submitted for possible publication. "The teams found that the odds of acceptance were twice as high for manuscripts for which authors had excluded reviewers compared to those whose authors had not done so."

This argument—that research reports end up in the hands of people most likely to endorse them—is one of the arguments climate deniers use whenever anyone points out that, according to the overwhelming weight of peer-reviewed evidence, they are wrong to dismiss the idea that human activity is causing climate change. Creationists make the same argument. They say biology journals won't publish their papers challenging the theory of evolution because they are biased against God in favor of Darwin. In the case of climate deniers and creationists, the scientific establishment replies that these groups do not produce the kind of work that passes scientific muster, regardless of its conclusions.

The creationist group Answers in Genesis, proprietors of the so-called creation museum that opened in 2007 in Petersburg, Kentucky, got around this problem by establishing its own journal, *Answers Research Journal.* According to a call for papers posted on its website when the group announced the journal's launch, "this will be a professional peer-reviewed technical journal for the publication of interdisciplinary scientific and other relevant research from the perspective of the recent Creation and the global Flood within the Biblical framework."[43]

In April 2016, the American Association for the Advancement of Science (of which I am a Fellow) ran a forum on still another problem in peer review: bias based on gender, ethnicity, disability, nationality, and institutional affiliation. Participants cited studies showing that women and African American researchers are held to higher standards. And they recommended a number of steps to improve things, including training reviewers and double- or even triple-blinding the review process so that even a journal's editors do not know who is the author of a paper under review.[44]

"Bias in peer review rots the scientific enterprise from within," Rep. Jackie Speier, a Democrat from California, told the conference. "We need the best, the most creative ideas to rise to the top."[45]

So—where does all this leave us? View with suspicion any claim made without the benefit of peer review. At the *Times* we do not necessarily ignore such work, but we want to know why it has not been reviewed. Remember, though, that the fact that work has been peer reviewed guarantees little.

People perennially propose improvements to the peer review process. For example, in 2015 the journal *Nature* announced that it would allow would-be authors to keep their names and affiliations off their submissions when they are sent out for review. (Given the degree to which people specialize in scientific research, it is far from clear that merely removing an author's name would in practice render a paper anonymous.) Others have proposed disclosing the identities of all involved, authors and reviewers alike.[46]

Others say researchers should be made to repeat their experiments—successfully—before being allowed to publish.

Still others advocate posting papers to public access "preprint" sites, where people can read papers even as scholarly journals review them. One major site is arXiv.org, a kind of electronic bulletin board used primarily by computer scientists, mathematicians, and physicists. Some journals are likely to object, especially if their business model relies on the exclusive access to hot papers they offer. Many of them have long refused to publish any paper whose contents have substantially been made public before publication in the journal. For many researchers, that policy is a serious disincentive.

For the most part, researchers given the chance to choose are sticking with the existing system.

One change that may take hold more widely is the practice of posting one's data online and inviting comment. Then, as the

Economist noted, "anyone can appoint himself a peer and criticize work that has entered the public domain."[47]

That's what happened when researchers from the Harvard-Smithsonian Center for Astrophysics studying cosmic background radiation—the celestial echo of the Big Bang—discovered what they believed was the signature of primordial ripples in the fabric of space-time, or gravitational waves. The discovery was significant because it seemed to confirm an important theory, called inflation, that helps explain why the universe developed into what we know today.

They made their data available online, prompting hundreds of physicists to check their work. Some of them challenged it, saying the data were unreliable. The result, as the European Space Agency tactfully suggested on its website in 2015, is that evidence of gravitational waves remains "elusive."[48]

Gravitational waves were the subject of a peer-review contretemps famous in the annals of science. It involved Albert Einstein, who with his collaborator Nathan Rosen submitted a paper in 1936 to the journal *Physical Review*. Their subject was gravitational waves, and they concluded that these waves did not exist. The paper received a cool reception by its reviewer, and when Einstein found out about that he was furious. He withdrew the paper and told the journal, "I see no reason to address the—in any case erroneous—comments of your anonymous expert."[49]

The reviewer, later identified as Howard Percy Robertson of Princeton, approached Einstein afterward to explain his thinking; the paper was rewritten to say gravitational waves might exist.

In February 2016, an international team of scientists called the Laser Interferometer Gravitational-Wave Observatory, or LIGO, announced that observations of two black holes colliding a billion light-years from Earth had confirmed that the waves did exist.[50]

But Einstein never submitted another paper to *Physical Review*.

3

THINGS GO WRONG

Misconduct

The history of science is full of error and even, at times, fraud. All human enterprises are full of error and even, at times, fraud. Ptolemy is said to have copied star charts from an earlier astronomer, Hipparchus of Rhodes.[1] The influential British IQ researcher Cyril Burt reported nonexistent studies and ascribed some of his findings to nonexistent collaborators. Researchers have achieved desired results by drawing on mice with felt-tip pens.

Still, compared to many other fields of human endeavor, research is relatively free of misconduct. Seventy years ago, one of the leading theorists of how science works offered an explanation of why that was so. He was Robert Merton, a professor of sociology at Columbia University, and his field was the sociology of science. In books and essays he described what he called "the ethos of science," the norms of behavior that, he wrote, had guided the field from the days of Newton.

He was not making them up, or writing about what he thought these patterns of behavior *ought* to be. He was describing what they were, putting them into a sociologist's framework that became a landmark in the philosophy and history of science.

The goal of science, he wrote, was the accumulation and sharing of what he called "certified knowledge"—empirical evidence, adequate and reliable, that could be used to make predictions and test them. These inquiries were characterized by free and open exchange of findings, the "disinterested" pursuit of truth unmotivated by anything other than the desire for knowledge, and reliance on nature—not culture, religion, economics or politics—as the final arbiter.

These values were institutionalized in ways held to be binding:

- prescriptions or requirements for certain actions,
- proscriptions barring certain unacceptable actions,
- preferences for some actions over others, and
- permissions.

According to Merton, all of these traditions of action are transmitted by precept and example. In other words, young researchers learn them in the lab or in the field, working at the elbows of their mentors. The result, he said, was a kind of scientific conscience or superego.[2]

And the result of *that,* he went on, was the virtual absence of fraud in science, a record he said "appears exceptional when compared with the record of other spheres of activity." That was true when Merton wrote it in 1942, and in my opinion it is true today.

Researchers often attribute this good record to some sort of innate moral superiority among those drawn to a life of research. I doubt they are right about that. As Merton said, there is "no satisfactory evidence that such is the case." A better explanation, he wrote, is the fact that science is self-correcting; that is, "the activities of science are subject to rigorous policing, to a degree perhaps unparalleled in any other field of activity."[3]

Still, the field is not absolutely free of misconduct, and responsible researchers worry that even a few instances leave the scientific enterprise as a whole under a cloud, not just among members of the public, but also among potential funders. There have long been efforts to promote what is called "research integrity," but as researchers who study the issue said in an essay in the journal *Nature,* "Misconduct continues, and evidence suggests that increasingly stressful competition for funds and the rush to publish may further erode ethical behavior."[4]

The erosion of scientific ethics begins in undergraduate years, they wrote, where "sharing" has become pervasive, and the kind of cutting and pasting an ordinary person would describe as plagiarism has become almost normative.

Gerald Koocher, a psychologist at Simmons College, in Massachusetts, and Patricia Keith-Spiegel, a professor emerita of social sciences at Ball State University, in Indiana, addressed the issue in an online survey they reported in 2010. They said that of the 2,599 scientists who responded to their survey, 406 reported witnessing no wrongdoing in their labs. Although some of the incidents involved careless record-keeping or inadequate supervision of research assistants, other problems—reported by hundreds of respondents—included fabrication or falsification of data; questionable publication practices, including publishing under their own names material written by others; plagiarism; rigging experiments to produce desired outcomes, and other problems.[5]

In 28 percent of cases, respondents who reported challenging the misconduct found that offenders changed their ways, eliminating the problem. On the other hand, Koocher and Keith-Spiegel wrote, in 27 percent of cases, suspected offenders denied that a problem existed.

Maybe they were right. Maybe they were wrong, but fixed things quietly. But to me the most notable finding in the survey is the extent to which respondents said they have noted events or practices worthy of challenge. That finding is alarming because, as Merton wrote seventy years ago, graduate students and postdocs learn the ways of the lab world by observing the conduct of their advisers and other senior researchers.

Agencies like the National Science Foundation, the major federal funding agency for the physical sciences, now require applicants for research grants to pledge to instruct people working in their labs on what is called "responsible conduct of research." Many commentators suggest the requirement is honored in the breach. Be that as it may, it was graduate students who raised questions about the work of Marc Hauser, an evolutionary biologist who resigned from Harvard University in 2011. In 2012 the federal Office of

Research Integrity found he had engaged in misconduct. According to the journal *Science,* his published work included "numerous mismatches between submitted papers and raw data."[6]

As far as we know, out-and-out invention is rare. But with grants harder and harder to come by and big business a growing supporter of research funds, distorting results to appease grant-makers or commercial benefactors can seem like a survival instinct.

In a commentary in 2005 in the journal *Nature,* researchers at the University of Minnesota and elsewhere reported that a third of the researchers they surveyed had changed experiment designs under pressure from a funding source, overlooked questionable use of data, ignored contradictory data, and engaged in other questionable practices. The researchers said efforts to foster scientific integrity should focus on this kind of thing as well as on the so-called "big three" of research misconduct: fabrication, falsification, and plagiarism.[7]

A couple of months earlier, researchers at the Harvard School of Public Health and the University of Massachusetts, Boston, reported results of a survey of medical school researchers who test new drugs for pharmaceutical companies. They found medical schools varied widely in their standards of acceptable conduct of pharmaceutical research. Seventy percent of such research, the researchers said then, was supported by the industry. For example, some allowed drug companies to insert their own analysis into research reports, suggest revisions before reports were published, or even write them in the first place.[8]

The study involved a survey sent to 122 accredited medical schools in the United States. The researchers received responses from 107 institutions. Of those, more than half handled 100 or more clinical trial agreements per year. While the majority of these institutions reported standards or rules to prevent misconduct, a sizable minority did not. (And the researchers could not say what went on at the other fifteen medical schools.)

Among responding institutions, 89 percent would not allow industry sponsors to revise a research report, and 93 percent would not allow them to suppress publication of results. But 96 percent would allow sponsors to review work for a period of time before publication. And while two-thirds prohibit researchers from discussing an ongoing trial with its sponsors, two-thirds would allow such discussion after the trial ends but before its results are published.

Half would allow sponsors to draft reports on the research they sponsor; 40 percent would not; 24 percent would allow sponsors to insert their own statistical analysis into papers, 47 percent would not, and 29 percent were unsure. More than a third of institutions (41 percent) would allow sponsors to bar investigators from sharing results with third parties, a crucial part of research.

After the results of this survey were made public, I interviewed some of the researchers about them. They told me that they realized, too late, that there were some defects in their own study design. For example, changing a study's design in midstream because of safety concerns expressed by National Institutes of Health, the major U.S. funding agency for medical research, is not the same as altering a design to get a result a sponsor wants.

Nevertheless, evidence is accumulating that questionable practices are not uncommon, and that dealing with the problem will require more than stricter regulations. For one thing, that could lead to "venue shopping" by pharmaceutical companies and other private funders of academic research, who will send their research money to institutions most likely to do what the funders want.

In 2002, the Institute of Medicine, the medical arm of the National Academy (and now the National Academy of Medicine), looked into the issue and concluded that there were no established methods for assessing integrity in the research environment. Therefore, it said, there was not enough evidence to say that any particular approach was best—beyond the widely held view that the policies and procedures already in place are necessary—but inadequate.

What happens when someone reports research results and then it turns out that the findings were mistaken, or even made up? The paper or papers are retracted. That is, the journal that published them prints a notice—a retraction—telling readers that it no longer stands behind the research. Usually it explains why.

If they discover their report was wrong, responsible researchers notify the journals that published it, so it can be retracted. Sometimes, a reader or reviewer (or rival) notifies the journal of problems with a paper it has published and a retraction results.

If you are a researcher, having your paper retracted is at least a humiliation and at worst a calamity. Perhaps for that reason, journals usually move deliberately (that is, slowly) when a paper comes under challenge. Cases are often complex, and anyway the journal usually wants to see what the institution involved says about any problems with the research. Some journals ask institutions involved in charges of research misconduct to produce a report describing how they are investigating the situation.

In some major cases, problems with research reports turn up because someone notices that a researcher has used the same image or data has been wrongly used in several applications. In an editorial in 2010, the journal *Nature* appealed to readers "who think the images of other information have been inappropriately handled to bring your concerns to the attention of the editors."[9]

Sometimes retractions make big news, usually because the original claim made big news. The South Korean cloning claim was one example. Another was a report in *Science* in 2009 that a mouse leukemia virus was linked to chronic fatigue syndrome in people. That study had been hailed by advocates for people with the syndrome, because it seemed to offer the first concrete evidence that the disabling condition might have a physical rather than psychological origin.[10]

In 2011, the journal *Science* retracted the paper. Soon after, all seven authors of another paper that had appeared to support the original paper's findings retracted their report, which had appeared in 2010 in the *Proceedings of the National Academy of Sciences*. They said they have been unable to replicate their own results.[11]

Often, though, retractions draw little attention. That is why, in 2010, Adam Marcus, managing editor at *Anesthesiology News,* and Ivan Oransky, executive editor of Reuters Health, started the website Retraction Watch, which tracks retractions of scientific publications. At first, they told the journal *Nature,* they feared they would not be able to update the site regularly, for lack of material.[12] But, they continued, "in just 16 months we have written about some 250 retractions." They added that in the previous decade retractions had risen fifteen-fold, while the number of scientific papers published increased by less than 50 percent. According to research reported in the online journal *PLoS,* the rate of retractions rose by a factor of eleven between 2001 and 2010.[13]

In a study reported in 2012, researchers who studied more than 2,000 retractions found that about two-thirds involved fraud, suspected fraud, or plagiarism. Only about 20 percent involved researcher error.[14]

Inability to replicate published findings is common. Two notable recent examples involved a survey of cancer research reports (six out of fifty-three studies held up) and, notoriously, a study of psychology papers (only thirty-nine of 100 papers passed muster).[15]

In 2015, Federal Reserve analysts reported that the results of two-thirds of economic studies it analyzed—from peer-reviewed economics journals—could not be replicated.[16] Commentators suggested the findings mean the so-called "dismal science" is not so scientific after all. I think they show the field conforms all too well to the norms of science!

There are those who suggest that the most prestigious journals—journals with the highest impact—may actually end up reporting more work that turns out to be hyped, faulty, or flat-out wrong. Publication in a top-tier journal is so helpful in advancing a researcher's career that it would not be a surprise if people seeking publication pressed their scientific case too strongly. Also, of course, research reported in top tier journals attracts more attention than findings published elsewhere.

Still, considering how much research work is published, retractions are rare. Marcus and Oransky estimated that about 1.4 million papers are published each year, and only about 300 retractions. And they attribute the rise not to an increase in scientific misbehavior but, at least in part, to plagiarism-detection software and the fact that the internet results in "a larger number of eyeballs on papers."

But mechanisms to deal with retraction-worthy work are inadequate, Marcus and Oransky wrote in their *Nature* commentary. Letters to scientific journals that published the work can be effective, but journals have limited space for them, at least on paper, and their publication may be delayed. Researchers can make online comments on papers, they went on, but "hardly anyone" does—and, they add, most journals do not look into complaints made anonymously. They called for a more regular system of "post-publication review" of scientific papers.

In another essay highlighting weaknesses of the retraction system, Richard Van Noorden, an assistant news editor at *Nature,* noted that journal policies vary widely from journal to journal, and that the retraction process itself is something of a blunt instrument, tarring everything it touches with the suggestion of misconduct, even if the issue is one of record-keeping or authorship. This stigma, he wrote, "may keep researchers from coming forward to admit honest errors."[17]

One suggested remedy is a system is called CrossMark, a collaborative organized by publishers of for-profit and learned society journals.[18] In theory, it will enable readers to check any paper for corrections, revisions, or retractions.

Such a system might deal better with questions of nuance, technical authorship issues, and other problems that do not necessarily speak to a paper's research conclusions. It would certainly give publishers a tool other than the blunt instrument of retraction. Though this system might mean fewer retractions and therefore less material for Retraction Watch, "we would be happy," Marcus and Oransky wrote, "if it meant that the scientific record had become more self correcting."

It might also reduce another problem—the fact that researchers sometimes continue to cite papers even after they have been retracted. *Nature* described work by John Budd, of the School of Education at the University of Missouri in Columbia, who looked at 235 papers retracted from 1966–1996. He found the retracted papers were cited more than 2,000 times subsequently—and only 8 percent of the citations noted the retraction.[19]

"There is nothing more irritating than publishing a paper that completely disproves every major conclusion of a study, and then years later seeing reviews or other papers cite the original (wrong) study, without the authors being aware that any doubts were ever raised," Thomas DeCoursey, a biologist at Rush University Medical Center in Chicago, told *Nature*. (The journal noted that he had raised questions about a paper it published; it took *Nature* four years to retract it.)

But the retraction situation is not always simple, especially when the finding in question is one that was made long ago. The advent of online retraction police "has made the standardization of processes to address allegations more complex," according to an April 2016 editorial by Marcia McNutt, editor-in-chief of the journal *Science*.[20] Editors want to note error, she said, but they want to be fair to authors. And what about a statute of limitations? Dr. McNutt cited cases in which journals were asked to retract research publications more than fifty years old. "This clearly raises the question of due process," she wrote. "None of the authors were alive to respond to the charges of misconduct."

Science in Court

When science and the law meet in a courtroom, the results aren't always what we might want.

In the first place, there can be science-related prosecutorial incompetence or misconduct—the kind of thing investigators discovered when they looked at state crime labs in Texas, North Carolina, and elsewhere.

Second, witnesses may make greater claims for their science than it can support. That seems to be what happened in the case of Cameron Todd Willingham of Corsicana, Texas, who was executed in 2004 after being convicted of killing his children by setting the family house on fire. At trial, the prosecution offered laboratory evidence and testimony by a local arson expert, who told the court the lethal fire had been deliberately set.[21] Only later did much of this evidence come under challenge. When it was reexamined, many people came away convinced that the state of Texas had executed an innocent man.

An expert panel convened by the National Research Council, the research arm of the National Academy of Sciences, reported in 2009 that the nation's forensic science establishment needed a top-to-bottom overhaul.[22] It recommended a board to assess subjects ranging from eyewitness testimony to fingerprint analysis. The Department of Justice established the National Commission of Forensic Science in 2013.

Another problem is the degree to which judges or lawyers understand scientific or technical evidence—and whether they can explain it to the ultimate lay audience, the jurors. This issue is not new. C. W. Griffin, an engineer and critic of the "antiscientific attitude" of some judges, wrote about it while the O. J. Simpson civil trial was underway. An alternate juror had been dismissed for "nodding into oblivion" during presentation of DNA evidence. Griffin's conclusion: "Jurors may be not only incapable of understanding complex expert testimony; they are sometimes even incapable of listening to it."[23]

One especially difficult issue is eyewitness testimony. Study after study has found that juries find it convincing. Study after study has also found that, on the whole, it is unreliable—perhaps the least reliable evidence routinely offered in court. Still, the Supreme Court ruled in January 2012 that courts are not required to subject eyewitness testimony to unusual scrutiny, unless it was produced through police misconduct.[24]

The vote was 8–1. Justice Ruth Bader Ginsburg, writing for the majority, said that the potential unreliability of eyewitness testimony is not enough, by itself, to render it unfair, especially since it can be challenged by cross examination and judges can warn about it in their instructions to juries.

In my view, Justice Sonia Sotomayor's dissent was more convincing. She noted that eyewitness testimony is uniquely persuasive as far as juries are concerned, and at the same time highly unreliable and vulnerable to suggestion. She cited research suggesting that 76 percent of the first 250 wrongful convictions overturned by DNA evidence since 1989 "involved eyewitness misidentification."

In 2012 the state of New Jersey adopted new rules on the use of eyewitness testimony, including a requirement that jurors be instructed on how to assess it.[25]

Medical malpractice litigation is another area troubled by pervasive disconnects between science and law. Medical malpractice is a tort or injury, and when conservative lawmakers talk about tort reform, what they mean is finding ways to limit litigation like this. Malpractice torts are a rich source of money for plaintiffs and their lawyers.

Of course, there is such a thing as malpractice. Estimates vary, but it is widely believed that as many as 100,000 Americans die every year from the effects of medical malpractice. But researchers who study the issue have found that most malpractice litigation involves something else.[26] When researchers at the Harvard School of Public Health studied the issue they found the cases that ended up in court were, on the whole, not malpractice but rather instances in which patients did not do as well as they or their loved ones might have hoped. Meanwhile, the researchers identified many cases of what seemed to them to be obvious malpractice that never made it to a courtroom.

Science and the law do not mesh well in court, at least not in countries whose system is based on British common law. The disciplines approach the search for truth in fundamentally different ways. So the relationship between science and the law is almost inevitably problem-plagued.

Some of the difficulty arises from the way the researcher and the lawyer approach their work. In theory, each is looking to discover the truth. But when researchers look for truth they start by asking questions about nature, offering hypotheses to answer the questions, and testing the hypotheses through observation and experimentation. They regard all answers as provisional, standing only until they are successfully challenged. And they are duty bound to consider all but the most obviously misguided challenges.

When scientists and engineers look for information, they cast a wide net—and they would be unethical (never mind unreliable) if they ignored evidence simply because it did not line up with a favored hypotheses.

Lawyers know from the get-go what judgment they want and they seek information to build a case to support it. Unlike scientists and engineers, lawyers are under no obligation to mention anything that undermines their argument, unless they are prosecuting a criminal case. Lawyers can ask yes or no questions and demand yes or no answers; researchers dwell on caveats.

The situation is further complicated by a long-running debate about what kind of technical evidence is admissible in court, and from whom. Ordinarily, witnesses in a legal case can testify only to what they saw, heard, tasted, or otherwise experienced themselves. They may talk about facts, but they may not draw conclusions about them on the witness stand.

That is not the case with an "expert" witness like a scientist or engineer, whose role is to describe and explain the facts to the judge or jury. Often, these explanations can have big consequences in the courtroom; the verdict in a case may hang on what experts say about it on the witness stand. (That's why some lawyers describe a bit of persuasive scientific or engineering evidence—or a persuasive scientist or engineer—as a "meal ticket.") In that event, a judge's decisions as to who is qualified to speak, and about what, can virtually decide a case.

Questions about admissibility of evidence are particularly knotty when the scientific evidence is new. That situation arises surprisingly often, in part because for many issues serious research does not begin until someone starts litigating. (That was more or less the case with lawsuits over the safety of silicone breast implants.) In the real world decisions must often be made before the science on the question is well underway, when it is too soon to know if a claim is well grounded in science.

Until the beginning of the twentieth century, American judges generally evaluated the credentials of a would-be witness, not the substance of the testimony the witness might offer. That changed in 1923, when a federal court took up the admissibility of evidence obtained from a so-called "deception test," a precursor of the lie-detector (polygraph), which relied on measurements of systolic blood pressure.

In a widely quoted opinion, the Court of Appeals for the District of Columbia said: "Just when a scientific principle or discovery crosses the line between the experimental and the demonstrable stages is difficult to define. Somewhere in this twilight zone the evidential force of the principle must be recognized, and while courts will go a long way in admitting expert testimony deduced from a well-recognized scientific principle or discovery, the thing from which the deduction is made must be sufficiently established to have gained general acceptance in the particular field in which it belongs."[27]

Probably most of us would agree on the theoretical desirability of limiting evidence in court to that which is demonstrably true. But scientists do not always agree on what is demonstrably true. In any event, in the absence of anything better, the "general acceptance" standard set out in Frye prevailed for years.

In 1975, Congress codified new Federal Rules of Evidence. These rules allowed the admission of "relevant" evidence—that is, in the language of the law, evidence that has "any tendency to make the existence of any fact that is of consequence to the determination of the action more probable or less probable than it would be without the evidence."[28] When it comes to science and technology and other specialty subjects, the rules say later if "scientific, technical, or other specialized knowledge will assist the trier of fact to understand the evidence or to determine a fact in issue, a witness qualified as an expert by knowledge, skill, experience, training, or education, may testify thereto in the form of an opinion or otherwise."[29]

In other words, the characteristics of the would-be witness, rather than the potential testimony proffered, would decide whether a jury would hear it. There are those who say these rules flooded the nation's courtrooms with a torrent of "junk science." People with advanced degrees, affiliations with major universities or other seemingly legitimate credentials took the stand to promulgate erroneous ideas before judges and juries who lacked the technical knowledge needed to separate the wheat from the scientific or technical chaff.

In 1993, the question came to the U.S. Supreme Court in *Daubert v. Merrell Dow Pharmaceuticals, Inc.,* one of the most famous (critics would say infamous) cases in American scientific jurisprudence. The case involved two children born with birth defects. Their parents sued the pharmaceutical company Merrell Dow, asserting that the defects were caused by Bendectin, the company's prescription anti-nausea remedy. The children's mothers had taken the drug during pregnancy. Merrell Dow maintained that the drug was blameless, but meanwhile the number of suits against the company grew, to the point that Merrell Dow eventually withdrew the drug from the market.

But as Justice Harry Blackmun wrote in his majority decision, "a well credentialed expert on the risks from exposure to various chemical substances" had reviewed more than thirty published studies involving 130,000 patients who had taken the drug.[30] "No study had found Bendectin to be a human teratogen," he wrote, using a term for substances capable of causing birth defects.

The plaintiffs did not contest this characterization of the scientific record, Justice Blackmun went on. Instead, they brought their own experts who asserted, based on their own laboratory studies and animal tests, studies of the chemistry of the drug and "reanalysis" of the original studies, that Bendectin could cause birth defects.

The district court that heard the suit had been unimpressed. It ruled in favor of Merrell Dow, saying that scientific evidence is admissible only if it is based on scientific principles "sufficiently established to have general acceptance in the field to which it belongs." Given the "vast body of epidemiological data" that seemed to clear Bendectin as far as birth defects go, the evidence the plaintiffs sought to offer "is not admissible to establish causation."

Certainly the witnesses the Daubert plaintiffs wanted to put on the stand qualified as expert by the existing standards of the rules of evidence, even if their views about Bendectin were not generally accepted. The question was, should they be allowed to testify because they met the standard of the rules, or should they be barred from the witness stand because, in effect, their testimony could not pass the Frye test?

Justice Blackmun wrote for the court. "That the *Frye* test was displaced by the Rules of Evidence does not mean . . . that the Rules themselves place no limits on the admissibility of purportedly scientific evidence," he said. "Nor is the trial judge disabled from screening such evidence. To the contrary, under the Rules the trial judge must ensure that any and all scientific testimony or evidence admitted is not only relevant but reliable."

Experts may testify about their scientific or technical knowledge, Blackmun wrote, but only if their testimony is "scientific." More than subjective belief or unsupported speculation, it must be grounded "in the methods and procedures of science."

What might those be? The National Academy of Sciences, the nation's most eminent body of scientists, and the American Association for the Advancement of Science, a large organization of researchers, teachers and others, answered the question in a brief they filed as friends of the court. "Science is not an encyclopedic body of knowledge about the universe," they wrote. "Instead, it represents a *process* for proposing and refining theoretical explanations about the world that are subject to further testing and refinement." For testimony to be scientific, in other words, it must be derived by this scientific method and appropriately supported based on what is known.

Justice Blackmun elaborated on this standard in his opinion, setting out these requirements:[31]

- Fit. The evidence must be sufficiently tied to the facts of the case to help the jury resolve factual disputes.
- Falsifiability. Citing Popper, Justice Blackmun said any statements constituting a scientific explanation "must be capable of empirical test. . . . The criterion of the scientific status of a theory is its falsifiability, or refutability, or testability."
- Peer review. While some ideas may be too new or of too limited interest to have been published in peer-reviewed literature, "submission to the scrutiny of the scientific community is a component of 'good science,'" Justice Blackmun wrote, "in part because it increases the likelihood that substantive flaws in methodology will be detected." As a result, whether the evidence in question has been subjected to peer review is something judges should take into account in deciding whether to admit it at trial.
- Error rates. In considering evidence derived from particular techniques, judges should weigh what is known about the error rates of those techniques.
- Finally, there is the general acceptability standard of Frye. "Widespread acceptance can be an important factor in ruling particular evidence admissible," Justice Blackmun wrote. Or to put it another way, if a theory or line of reasoning is known and yet has attracted little support, it "may properly be viewed with skepticism."

By the way, Chief Justice William Rehnquist's dissenting opinion in Daubert offered, I assume inadvertently, a stunning demonstration of Snow's "Two Cultures" syndrome at work. He was at a loss to understand what scientists could mean by "falsifiability," Rehnquist wrote. Given the importance of the concept in the scientific method, this assertion, by the man who was then the most prominent member of the nation's judiciary, is alarming. It certainly calls into question Justice Blackmun's optimism that trial judges are capable of determining whether or not a theory is falsifiable.

In the years since, other scholars and judges offered elaborations on the meaning of Daubert. Some, for example, advised that judges differentiate between research findings developed independently of the issue at hand and research performed to produce evidence for use at trial—something some commentators call "litigation science."

But at a conference on the subject in 2006, other experts noted that research directed at producing admissible evidence is not necessarily invalid, and that research conducted purely to increase the sum of human knowledge is not necessarily free of error or bias.[32]

Two subsequent cases, *General Electric Company v. Joyner* and *Khumo Tire Company v. Carmichael,* expanded the court's guidance in Daubert. In Joyner, judges were empowered to reject expert testimony if they distrusted the connection between witnesses' data and the conclusions they drew from the data. Kuhmo expanded Daubert and Joyner to cover any kind of "technical" information.

In the wake of Daubert, the government established the Science and Technology Resource Center at the Federal Judicial Center in the Department of Justice, which, in 1994, also published the first edition of the *Reference Manual on Scientific Evidence,* meant to serve as a rough guide for judges dealing with scientific and technical issues. The first issue contained guidance on the admissibility of expert evidence, seven areas of expert testimony, and the use of special masters. In 2000 the Judicial Center issued a second edition with information on how science works, medical testimony, and civil engineering. The center has also produced video programs on scientific evidence.

Meanwhile, some judges, alarmed over the technical testimony they find themselves called upon to assess, have been taking things into their own hands, organizing seminars on topics like DNA evidence. The Brooklyn Law School has had "science for judges" programs in which participants heard presentations on such topics as epidemiology, toxicology, and statistics—the issue "most troublesome for the attendees."[33]

From time to time, people have proposed creating quasi-judicial "science courts" to resolve scientific and engineering disputes before they get to the witness stand. Generally, lawyers view these proposals with suspicion, in part because they fear giving so much power and influence over their juries to a supposedly authoritative entity over which they have no control. Scholars like Sheila Jasanoff, a lawyer and professor of science and technology studies at Harvard, have described the idea as "unworkable."[34]

Still, there is growing agitation for something, given what remains a substantial knowledge gap between the nation's judiciary and its scientific and engineering community. This gap was vividly illustrated in 2006, when the Supreme Court heard arguments in a climate-related suit brought by twelve states in addition to several cities and environmental groups. The suit charged that the Environmental Protection Agency was failing to regulate emissions of heat-trapping gases. At one point, when advocates for both sides were making oral arguments before the court and responding to questions from the justices, Justice Antonin Scalia was corrected on the issue of where in the atmosphere carbon dioxide accumulates.

"Troposphere, whatever," he said. "I told you before I am not a scientist."[35]

Researchers and Journalists

I joined the science department of the *New York Times* in 1986 and have worked there for most of the time since. I believe my colleagues in the science department are the best science news team of any lay-language news organization in the world.

Yet we struggle to stay on top of a flood of possible stories, to decide what we must cover and what we can pass up, to discern what is a genuinely important discovery and what is a flash in the pan or a trip down a trivial research byway. And we are playing at the top of the game. Science journalists elsewhere, talented and diligent though they may be, work under far more difficult conditions.

For one thing, in general journalists are as ignorant as anyone else, and we do our share of irrational thinking. We improve with experience, and journalists who are allowed to spend time on the science beat develop expertise. But in mainstream journalism, senior news executives are often suspicious of people who have been on any beat long enough to develop real knowledge—they worry about our "getting in bed" with our sources. These days, people employed by news organizations to spend full time on science coverage are harder and harder to find.

A number of programs exist to train journalists in one research field or another, to show them how researchers work and how to assess scientific or engineering claims. One is the Metcalf Institute for Environment and Marine Reporting, based at the Graduate School of Oceanography at the University of Rhode Island. The Metcalf Institute runs a five-day program each summer in which journalists spend time with researchers to see how they do their work. And it also offers lectures and other programs for journalists (and the public) around the country. Immersion programs for journalists also exist elsewhere. For example, the Knight Foundation sponsors weeklong bootcamps and yearlong programs at the Massachusetts Institute of Technology and elsewhere.

I was a founding member of the Metcalf advisory board, and I have always been inspired by the journalists who take their own vacation time to participate in our programs. But nowadays many would-be participants are so worried about their jobs that they are reluctant to request the time off. That is especially the case for the yearlong programs. In a shrinking industry, journalists fear that when they are ready to return to work, refreshed and filled with new knowledge, their jobs will have vanished.

As everyone knows by now, the internet is forcing big changes on the mainstream media, traditionally the major source of lay-language science news in the United States. Our business model is broken and we have not yet figured out how to fix it. After years of giving away content online, publications have embraced paywalls, but in most cases their revenue streams do not yet come close to the profits media companies used to glean from paper.

Of course, money problems in the newsroom are nothing new. When I began my career at the *Providence Journal,* there was an editor who began any meeting about any proposed project by writing a message on the conference room's white board: "There is not enough time, there is not enough staff, there is not enough space." In other words, there was not enough money. Then we would discuss how we would do what needed to be done with the resources we had.

According to media analysts of all stripes, conditions are far, far worse today, almost across the board. In a presentation at Harvard a few years ago, Tom Rosenstiel, then director of the Project for Excellence in Journalism, said that in 1988 the average television reporter's workload was 1.3 stories per day.[36] By now reporters are routinely expected to report two stories per day. Stories are also getting shorter. At one station in Phoenix, Rosenstiel said, the average story lasted just eleven seconds. That's longer than you think—try counting it out—but far less time than a responsible journalist needs to cover anything complex, like science or engineering.

According to Jeff Burnside, an Emmy Award–winning environment reporter for NBC affiliates in Miami and Seattle and now the president of the Society of Environmental Journalists, after promos, commercials, weather reports, and sports, the average half-hour newscast contains only about eighteen minutes for everything else. Even a good science story must fight hard for a place in that space.[37]

The Weather Man and Climate

A few years ago, I attended the National Hurricane Conference, an annual meeting for researchers, public officials, meteorologists, and others concerned about hurricane safety and related issues. The meeting was held in Nassau, and I rode from the airport to the headquarters hotel in a shuttle bus with another attendee, a television weatherman.

On the way, he spoke at length about human-induced climate change, an idea he dismissed as nonsense. I kept silent—though I wondered how someone so ludicrously uninformed about climate could work as a weather reporter.

A couple of days later I had a bigger shock. After a panel discussion on hurricane prevalence, one member of the audience, also a television meteorologist, rose to challenge the assertion that the burning of fossil fuels was warming the earth. He called the idea "a communist plot." The audience erupted in wild applause.

As it turns out the community of television meteorologists has been a center of climate denial. Some attribute it to the fact that their weather predictions so often turn out to be inaccurate. In any event, many of the nation's weather reporters have lost faith in the models of climate generally. Eventually, the American Meteorological Society created a "Committee to Improve Climate Change Communication" and commissioned researchers to "explore and, to the extent possible, resolve" tensions on the issue. In July 2014 the society reported the results in an article in the *Bulletin of the American Meteorological Society.*[1]

The researchers, from George Mason University and elsewhere, found that among meteorologists surveyed, those with climate expertise, liberal politics, and knowledge that climate experts agreed on the issue were more likely to embrace the idea that human-induced climate change is real and potentially harmful. Those who believed climate experts disagree on the issue were more likely to be deniers. The researchers' advice to the meteorological organization: "convey the widespread scientific agreement about climate change; acknowledge and explore the uncomfortable fact that political ideology influences the climate change views of meteorology professionals; refute the idea that those who do hold nonmajority views just need to be "educated" about climate change; and continue to deal with the conflict among members of the meteorology community."

Did the work have any effect? It's hard to say. Many meteorologists persist in climate denial. I might have been tempted to dismiss this controversy as one with few implications in real life, except for this alarming fact: in most of the nation's local television news operations, the only journalist with any scientific responsibilities or expertise is the weather reporter.

---

1 Neil Stenhouse et al., "Meteorologists' Views About Global Warming: A Survey of American Meteorological Society Professional Members," *Bulletin of the American Meteorological Society* (July 2014); http://journals.ametsoc.org/doi/abs/10.1175/BAMS-D-13-00091.1.

Surely, you may be thinking, things must be better for newspaper journalists. There was a boom in newspaper science reporting in the 1980s, with freestanding science sections blossoming at papers from the *Boston Globe* to the *Los Angeles Times*. By now many of them have died or been drastically pruned. Even CNN has abolished its environmental news team.

The situation is further complicated by a widespread reluctance among scientists and engineers to engage with the public or journalists. When I speak to researchers, which I do fairly often, my message is this: coverage of scientific and technical subjects in general is not as good as it should be but, especially in the current media environment, researchers are the only ones who can make it better, by agreeing to speak with reporters and putting in the time and effort it takes to make things clear to us.

Too often, though, talking to reporters is a no-win proposition for researchers. They communicate their findings in scholarly journals, not in lay-language news reports. Decisions on whether they will be tenured, promoted, or awarded research grants do not normally hang on what appears in the media.

If researchers are in the newspaper or on television or radio too much—and their colleagues may set that bar rather low—they become known among their colleagues as publicity hounds or polemicists who have abandoned the purity of the laboratory for a life of celebrity. Some scientists call this "the Carl Sagan effect," after the Cornell University astronomer whose series for public television, *Cosmos,* was highly successful. The more it was praised, it seemed, the more other scientists criticized him for the time he spent away from his research. As a result, he was denied membership in the National Academy of Sciences. (In what I can only assume must have been a gesture of remorse, the Academy honored him after he died.)

Also, when scientists do cooperate with journalists and take the time to describe their work, all too often they find themselves quoted in a report that is shoddy, inaccurate, or hyped. Pushy, unprincipled, ignorant, and shallow—those are some of the milder epithets I have heard scientists and engineers apply to me and my fellow practitioners.

As a journalist, I have to concede there is some truth in the scientists' indictment. In large part, though, errors result because science journalists don't always get the help they need to get things right. Often scientists are unable or, worse, unwilling to tell the story in words a lay audience can understand. (Sometimes they cannot even tell the story in words their fellow scientists can understand. As my *New York Times* colleague James Glanz once put it, apropos of physics journals, "the flood of unexplained acronyms, cryptic symbols, endless sentences, and nightmarish graphs has risen so high, say some leaders in the field, that physicists can no longer understand each other.") And Jim has a doctorate in physics![38]

As a result of all this, many researchers regard time spent with journalists as time wasted, and journalists regard scientists and engineers as elitist, unable to talk except in jargon, obsessed with trivial details, isolated in ivory towers, and unwilling to take a stand on matters of public importance.

It is sad that this is so. Journalists and scientists have much in common. Both groups are curious, analytical, skeptical, competitive, highly motivated, resistant of authority, and self-centered. Also, both groups delight in learning things and then telling other people about them. But researchers and journalists view the world through different prisms and seek different kinds of information from the processes of science.

Researchers approach questions rationally, in orderly fashion and in accord with widely accepted laws and methods. The public arena, where journalists live, is ruled by emotion, prejudice, *post-hoc-ergo-propter-hoc* thinking, and other forms of unreason.

Scientists and journalists tell stories differently. In a typical research paper, scientists describe the question they hoped to answer, the methods they used to gather data about it, the data themselves, and the conclusions they draw from them. Journalists tell this story in reverse.

Scientists are used to counterintuitive findings, statistical anomalies, and so on. They reason probabilistically. They understand that a fair coin flipped five times is as likely to come up HHHHH as it is HHTHT. Like the general public, journalists tend to hold to what is intuitively obvious even in the face of clear evidence that it is wrong.

Scientists deal in probabilities; journalists seek certainties. They are not comfortable with odds ratios and error bars. Everyone who has been a reporter has heard an editor bark, "Don't waste my time with something that MIGHT be true, tell me what IS true!" The researcher is mystified by this instruction.

Journalists want answers quickly. They hate to have to watch the agonizing slowness with which experiments are repeated and repeated as evidence accumulates until an idea wins scientific acceptance. Scientists' insistence on deliberate, steady progress toward statistical significance can render their work moot, in that by the time they are comfortable making pronouncements they may be bringing up the rear. This problem is particularly troublesome with environmental research, which tends to move incrementally. "You'll never wake up to a headline "Global Warming Breaks Out," my *Times* colleague, the environment blogger Andrew C. Revkin, often says. "Journalism doesn't notice things that happen over a hundred years."

Journalists look for sweeping claims; scientists shy away from them. Responsible scientists are not particularly eager to announce that this or that question has been settled. Perhaps because of the growing difficulties of obtaining and maintaining research grants, they emphasize what remains to be learned at the expense of what is known.

Scientists explore things because they are interesting even if they have no immediate practical importance. We journalists need what we call "a peg"—an answer to the reader's question, "Why are you telling me this—and why now?" A related point: before they can sell a story to an audience, reporters must first convince skeptical editors or news directors that a story is worth doing. Sometimes that pressure can cause journalists to drift into hype.

Many scientists believe journalists who specialize in science ought to be advocates for science. At one time, science writers apparently thought so too; the National Association of Science Writers included that item in its agenda when it was organized in the 1930s. Today, responsible journalists regard scientists and engineers with the same skepticism they bring to other news sources. Researchers can find that offensive.

Scientists subject their work to peer review. They do not understand why journalists do not adopt a similar approach. That is, they do not understand why journalists will not send them drafts of articles for their review before they are published. Journalists explain that it would be unwise, to say the least, to allow news sources to review or approve copy. Scientists reply that they are not ordinary news sources, they are seekers of truth, far too high minded to interfere with copy to advance their own ends. When journalists raise an eyebrow, scientists take offense. As covering research gets harder and harder, more and more science journalists say they will run their copy by their research sources. I don't like the idea, but I can see why other journalists find it tempting.

Many science and engineering stories are complex. Journalists tend to avoid complex stories; they embrace a kind of self-imposed censorship, avoiding stories they fear they won't be able to tell well in the short time they have to learn the facts and report them to viewers, listeners, or readers. Or they worry that if they do take on such a story, they will end up with something that emphasizes the views of people who cooperate with them—who take their calls and answer their questions—even if the situation overall is more complex. Or they may rely overmuch on official sources, especially government agencies with efficient public relations operations, or even certain advocacy groups. Journalists may believe that the pronouncements of these organizations have more weight and are therefore more objective. They take refuge behind official data.

Even big-name news outlets may produce reports that don't hold up. One of my favorite examples was an eruption of coverage of what was described as "a study by experts in Germany" that supposedly showed that people with blonde hair are an endangered species that would be extinct by 2202. (The last natural blonde would be born in Finland, the "study" supposedly found.)[39] Journalism organizations ranging from the BBC to CBS, ABC, and CNN picked up the story, interviewing dermatologists, experts on genetics, and others who offered their own opinions on the supposed finding, which was sometimes described as coming from the World Health Organization. What too many journalists did not do, unfortunately, was check with the WHO, which eventually issued a statement saying it had never conducted any such research.[40]

This kind of news item is what we in journalism jokingly call "a story too good to check," because it is almost certain to unravel under even tepid inquiry. One of my *New York Times* colleagues used to grumble regularly that he had "reported a story off Page One"—that is, he had uncovered enough conflicting data and unanswered questions to turn a story filled with exciting flat assertions into one filled with nuance and, possibly, confusion. It is no fun when your assiduous reporting turns your potential big story into a ho-hum item—but it is the mark of the responsible journalist.

Meanwhile, in our increasingly partisan and bitter political climate many in the media find themselves trying harder and harder to be fair to all sides—even when a story does not actually have more than one side. For conventional journalists, this need for "objectivity" is intuitively obvious. But in science writing it carries unusual risks. Reporters eager to avoid tendentiousness or editorializing—that is, eager to give both sides—can wind up in a kind of analytical passivity that leaves them with he-said-she-said stories that simply shift the burden of interpretation to readers, who are even less equipped for the job than they are.

In the process, we sow confusion on issues on which science is really more or less settled. I made this point in a commentary in the *New York Times* in a special issue marking the twenty-fifth anniversary of Science Times, its weekly section on science, medicine, and health.[41] The essay provoked a storm of emails from outraged researchers.

"There really is no serious science debate today in areas such as the reality of global warming, the basic mechanisms of biological evolution, the age of the earth, the expansion of the universe, the supposed harmful effects of cell phones and other sources of microwave radiation, the claims for health benefits of magnets, or the proposition that alien interstellar microbes are raining down on Earth," one of them wrote. "Yet in every case it is possible to find someone who will be glad to espouse the losing side in return for being quoted in the press."

Another wrote in an email that when a respected news organization pits members of the NASA Mars Global Surveyor team against people who claim an artificial face has been constructed on the planet, "one begins to lose all hope." "Quackulent quotations," a physicist called the phenomenon, in still another email.

Aaron Sorkin, the creator of the HBO series *The Newsroom,* got at this issue in 2012 in an interview in *New York Magazine.* "The thing I worry about is the media's bias toward fairness," he said. "Nobody uses the word 'lie' any more. Suddenly everything is 'a difference of opinion.' If the entire House Republican caucus were to walk onto the floor one day and say, 'The Earth is flat,' the headline on the *New York Times* the next day would read 'Democrats and Republicans Can't Agree on Shape of Earth.' I don't believe the truth always lies in the middle. I don't believe there are two sides to every argument. I think the facts are the center. And watching the news abandon the facts in favor of 'fairness' is what's troubling me."[42] He exaggerates, but not by much.

To journalists like me, mainstream or even "legacy" journalists, it seems fair to give all sides, even if some are outliers and a story really only has one side. Plus, sometimes the outliers turn out to be right. Certainly the mainstream has been wrong often enough that journalists are right to be wary of assuming that anything is true simply because a majority of scientists or engineers believe it.

Unless the journalist is an expert, and no journalist can possibly be an expert on everything she is called upon to cover, she has no way of knowing for sure whose views must be heeded and whose can safely be ignored. Even expert guidance sometimes fails. When I headed the Science Department at the *Times,* scientists would periodically advise me to hire staff with PhDs. My reply was always the same: "what in?" Even they do not know enough.

In the case of climate, for example, mainstream scientists say mainstream journalists for far too long gave too much coverage to those who profess to doubt that human activity is contributing to potentially disastrous changes in the climate, or who argue that nothing is changing, or who insist that any changes will be beneficial (and if they aren't, technological geniuses will engineer a fix). By now, the opinions of these climate deniers are notable more for the cover they give politicians than their scientific rigor. Today we journalists pay them less heed—but we do not ignore them.

Perhaps the most flamboyant example of this phenomenon that I ever encountered involved Clonaid, an organization that described itself online as the world's leading provider of human cloning services. In 2002, it announced it had successfully cloned a human being. The claim itself was not beyond the pale. Though scientists had long assumed cloning mammals would be impossible, Scottish researchers had cloned Dolly the sheep in 1997, and she was followed by cloned mice, cats, goats, pigs, and cows. Though no primates had ever been cloned, some researchers said techniques developed in fertility clinics might actually make that process easier, not more difficult, than cloning another kind of mammal.

Clonaid was unusual because it was affiliated with the Raelians, a religious sect based in France whose adherents maintained, among other things, that space travelers created the human race by cloning. At the *Times,* we ignored the announcement, on the grounds that it was obviously nonsense. But as a media furor grew, we eventually felt we had to weigh in, and we produced several articles and at least one editorial, all of which described the enterprise in highly skeptical terms—the editorial said the whole business was "worthy of a P.T. Barnum sideshow."[43] The story died out when Dr. Michael A. Guillen, a former science editor for ABC News enlisted by Clonaid to test the supposed clone and her supposed mother, withdrew from the project, saying it might be "an elaborate hoax."

But it was not the last human cloning hoax. Only a few years later Woo-Suk Hwang, a South Korean researcher, announced that he had produced cloned human embryos. There was nothing particularly suspicious about it, even though it was announced in a press conference at a scientific meeting, not in a peer-reviewed research paper—and I was among the people who contributed to coverage of it in the *Times.* That claim turned out to be a hoax as well.

In retrospect I don't see how we could have ignored either claim. But, especially in the case of Clonaid, I still worry about whether the mere fact that we paid attention suggested to our readers that the claim might have merit.

The fragmenting of the media audience and the weakening of the authority of "legacy" news outlets produces a kind of free-for-all in which, as my former *Times* colleague Todd S. Purdum wrote in an article in *Vanity Fair,* "the viral communities of the Internet make outright falsehoods nearly impossible to extinguish."[44]

He cited a *New York Times* / CBS poll that found almost half of Americans think President Obama was not American-born (Donald Trump, the most prominent "birther," did not back off on the claim until the summer of 2016), and a Harris poll that found that 57 percent of Republicans believe he is a Muslim. And, Purdum wrote, "for good measure 38 percent [of Republicans] believe he is 'doing many of the things Hitler did,' and 24 percent believe that Obama 'may be the anti-Christ.'" This too is nonsense. But, as Purdum went on to say, "in pursuit of objectivity, most honest news outlets draw the line at saying flatly that something or other is untrue, even when it plainly is."

Will the internet make things better? There are those who believe it will, that the blogosphere's embrace of "transparency" will
produce a marketplace of ideas in which good (reliable) information will win out. The rise of fake news throws that into doubt.

I fear for a society in which the discipline of verification is
replaced by the internet's journalism of assertion. In a prescient article in the *Washington Post* in 1996, Joel Achenbach talked about
what happens when people cannot tell good information from bad.
"The danger is that we are reaching a moment when nothing can be
said to be objectively true, when consensus about reality disappears,"
he wrote. "The Information Age could leave us with no information
at all, only assertions."[45]

In theory, sites like Wikipedia work on consensus, and when
there is consensus, Wikipedia works well. The journal *Nature* reported in 2006 that its entries were as accurate, on the whole, as
entries in the *Encyclopedia Britannica,* a publication driven out of
print by the internet. (Later it reported that the encyclopedia dis-
puted the claim.)[46] But as Adam Gopnik noted in the *New Yorker,*
"when there's easy agreement it's fine, and when there's widespread
disagreement on values or facts, as with, say, the origins of capitalism,
it's fine, too; you get both sides. The trouble comes when one side is
right and the other side is wrong and doesn't know it . . . Creationists
crowd cyberspace every bit as effectively as evolutionists, and extend
their minds just as fully. Our trouble is not the over-all absence of
smartness but the intractable power of pure stupidity, and no machine, or mind, seems extended enough to cure that."[47]

Gopnik's point is proven, by and large, by the commentary provided by readers that follows online news articles. Some news
organizations, including the *Times,* pay staffers to monitor online
comments to maintain some standards of decency and coherence. But
that's expensive, and some internet enthusiasts say it bleeds the brio
out of the online experience, leaving a kind of skim-milk commentary behind.

On the other hand, new research is beginning to suggest that the
stream of negative comments from readers that can accompany a

news report leads readers, independent of the facts, to assume there must be something wrong with the research. Beyond that, more and more information posted online comes from someone with an axe to grind—sometimes acknowledged, sometimes not. It would be nice to think that the rise of social media will mean a democratization of truth. But as long as genuine disagreement exists and as long as genuine venality and bigotry exist, I will not be holding my breath.

Meanwhile, it would be helpful for all concerned if researchers absorbed the message that it is important for them to know how to communicate in terms ordinary people can understand.

A major exponent of this view is the actor Alan Alda, who after years of playing Hawkeye Pierce on the television series M*A*S*H, among many other roles, became a host of several science programs on television, including *Scientific American Frontiers*. Among other things, he advocates teaching researchers the techniques of improvisation to encourage them to relate more naturally to their own information.

When it comes to communicating scientific or technical information, "you have to be trained and any natural talent has to be sought out and if you don't have talent you have to learn the journeyman techniques," Alda said. "But I believe it can be systematically taught." He added, "the Stanislavsky method taught even untalented people to perform adequately."[48]

Among other things, he thinks researchers should consider using metaphors and other figures of speech to explain their own work. "I really do not want half-baked metaphors," says Alda, who has interviewed scientists often. "I want metaphors that help me understand the real thing. And if possible I would like them to get me to understand it the way they understand it in the language of science—math and science."

Like others who attempt to explain science to the lay public, though, Alda finds that explanations often come surrounded by useless detail. Scientists who are good at explaining "are able to let you know they are leaving something out and for now you can leave it out but if you probe deeper you will need to know," he said.

All too often, though, researchers dismiss that practice as over-simplifying. But, as Alda put it, "Clarity is not the same thing as dumbing down." Alda, who commissioned and starred in a play about the physicist and Nobel Laureate Richard Feynman, noted that Feynman used to say "if you can't explain it to a child you don't really understand it." Alda was an organizer of a new program to train scientists and health professionals in speaking more clearly to the public—now named Alan Alda Center for Communicating Science, at Stony Brook University in New York. In 2016 he was honored with the Public Welfare Medal from the National Academy of Sciences.

Researchers do not devote much energy to this kind of thing; indeed, many of them think it is something to avoid. They believe their job is at the laboratory bench, period. ("Every moment away from the bench is time wasted," one climate scientist once told me.) Plus, researchers concerned about departmental budgets or grant money, which increasingly comes from private sources, fear speaking out on controversial topics. Young researchers, still working for tenure and promotion, are even more vulnerable. "It is disheartening to read that many scientists refuse to talk to the press," a researcher wrote me by email after my twenty-fifth anniversary commentary appeared. "If the good scientists retreat into their labs and lock the door, the vacuum will be filled by others perhaps less competent."

Where does all this leave us? I am tempted to offer the old adage, "don't believe anything you read and only half of what you see." Things are not quite that bad. But realize you cannot accept at face value anything you encounter in newspapers, online, or on broadcasts.

Pay attention to the source. Does the news outlet have a track record of reliability? Is it even for real? If not, ask yourself why you are bothering with it.

Seek journalism outside of your comfort zone. If you usually read the *New York Times* editorial page, try the *Wall Street Journal*'s. If you usually read the conservative journal *American Spectator,* try the *Nation.* If you usually get your news from public television or National Public Radio—well, I'd love to say try Fox News, but its record of inaccuracy makes me hesitate.

Beware of stories about "trends." We sometimes joked at the *Times* that if we could find three examples of something—people wearing baseball hats backwards, women quitting high-profile jobs to stay home with their children—we could call it a trend. But trends are more than collections of anecdotes. And remember, the plural of "anecdote" is not "evidence." If someone describes something as a trend, look for the evidence he uses to back up that assertion.

Be aware that reporters often describe people *we* believe as "expert" and organizations *we* trust as "non-partisan." Notice how the journalist ends her report. Often we give the last word to the person we believe is right. Be skeptical of us.

Unfortunately, people can make money generating clickbait, no matter how phony. So it proliferates. But there are a few things to check if you think you may be looking at fake news.

Check the website address given for the news outlet. For example, I was once surprised to see a respectable American news outlet reporting that there was a new focus on vaccines as the cause of autism. Only when I looked did I realize the report was actually coming from a phony site in Palau. Fake news, for sure.

Look at the date of the article—often promulgators of clickbait spread outdated stories. Get onto a site's home page and look for the tab "About Us," or the like. Look it up in Wikipedia. Outrageous claims and lots of words in capitals suggest you may be looking at fake news.

It is good to look beyond your comfort zone for news and information. But if you stumble across an extraodinary claim, check it out with news outlets you trust. Remember the adage of research: extraordinary claims require extraordinary proof.

4

THE UNIVERSAL SOLVENT

A Matter of Money

In 1955, as Jonas Salk's polio vaccine was starting to rid the world of a crippling and often lethal disease, the researcher was interviewed on television by the legendary broadcast journalist Edward R. Murrow. Murrow asked him, "Who holds the patent on the vaccine?"

"Why the people, I would say," Salk famously replied. "There is no patent. Could you patent the sun?"[1]

There are those who say Salk did not patent the vaccine because he couldn't, at least by the standards of the day, in large part because the vaccine relied too much on previous work by others. But many people were inspired by his assertion that rights to a vaccine against a dread disease should belong to "the people."

Today, the idea is laughable. No one takes a step in the lab (hardly) without filing a patent application. People argue about how this new research landscape benefits or hurts us. But no one can deny research is fundamentally altered under the influence of money. To see how, one need only read Robert Merton's mid-twentieth-century essays on the sociology of science.

In "The Ethos of Science," he described four norms he said prevailed in science—universalism, communalism, disinterestedness, and organized skepticism. Note that Merton was not recommending them, he was describing them—as they existed at the time. He called them the organizing principles of scientific research.

It's interesting to consider them now—and how much each of them has changed.

UNIVERSALISM

Universalism, as applied to science, means that the enterprise transcends culture, and contributions are accepted from all. Merton acknowledged that bigotry in science "is part of the historical record."[2] And as a Jew writing in 1942, after Hitler had purged Germany of "Jewish science," he had to know the issue had not been put to rest. Still, he wrote, in science the personal or social attributes of the scientist are irrelevant. He quoted Pasteur: "Le savant a une patrie, la science n'en a pas."[3] (Merton takes no notice of the virtually total exclusion of women and people of color from serious work in science, perhaps because they were excluded from so much in the 1940s.)

COMMUNALISM

The fruits of scientific investigation belong to all, Merton wrote. Originally he called this idea "communism"; later he changed his term to "communalism." As he put it, "the substantive findings of science are a product of a social collaboration and are assigned to the community. They constitute a common heritage in which the equity of the individual producer is severely limited." Scientists' claims are limited to public recognition and the esteem of their fellow researchers, he said. If the institution of science functions efficiently, people will receive recognition and esteem roughly in accord with "the significance of the increments brought to the common fund of knowledge."

People don't possess scientific information exclusively, even if they have discovered it, Merton added. "[P]roperty rights of science are whittled down to the bare minimum."[4] (In other words, "Could you patent the sun?")

Today, it is hard to believe that such a system could ever have been in place, but it was. And even when the profit motive began to appear, it did not at first reach far into academia, where much basic research is performed. Not so very long ago, academic researchers who wanted to commercialize their discoveries might form private companies to exploit them—but they would leave academia and enter the world of business to do it.

Today many universities have a special staff—usually called something like a "technology transfer office"—to turn its researchers into profit centers.

Disinterestedness

Many people think "disinterest" means lack of interest, but strictly speaking, and in the Mertonian sense, "disinterest" means lack of motivation by financial or other personal considerations. As Merton put it, scientists work without consideration of personal gain, or allegiance to any cause or interest other than the truth. And, he added, the public nature of the enterprise, the necessity of scientists' findings being testable, and the ultimate accountability of scientists to their peers promote disinterestedness as a norm.

Organized Skepticism

The ethos of science requires detached scrutiny, Merton wrote, the suspension of judgment until the facts are in. As Merton put it, this idea is both a method and a mandate. "The scientific investigator does not preserve the cleavage between the sacred and the profane, between that which requires uncritical respect and that which can be objectively analysed," he wrote.

This world view is one reason why I object when people describe as "skeptics" those who challenge the well-established idea that human activity is altering Earth's climate. Skepticism is the default position of science.

Until relatively recently, these Mertonian norms—universalism, communalism, disinterestedness, and organized skepticism—were widely regarded as more or less inviolate. Today, much of what Merton wrote is out the window, particularly when it comes to money. In the United States, science has become industrialized—tied to the business of making money in more or less direct ways.

In part, that relates to changes in the way scientific research is financed in the United States.

Many people divide the history of research in this country into three eras, differing in part according to who provided money for research and how they doled it out.

In the first era, in the years leading up to World War II, the center of scientific gravity in the world was in Europe, not in the United States. The country did well—very well—applying discoveries made abroad, but with some notable exceptions, basic research was generally given somewhat short shrift.

The second age of U.S. research might be said to have begun when Hitler, in defiance of Merton's universalism, drove Jews and others out of European science. Many of his scientific exiles, including Einstein, ended up in the United States, along with others similarly driven from Italy (Enrico Fermi), Hungary (Leo Szilard), and elsewhere. Einstein, Fermi, and Szilard were among the scientists most crucial to the development of the atomic bomb, among other things. They and their fellow fugitives from fascism changed the world of American science.

This influx of talent, plus the vast efforts to construct the bomb, perfect radar, and eventually rockets, computers, and other miracles of wartime engineering, created a strong new focus on research and, when the war was won, a growing appreciation of the potential fruits of research conducted for the sake of the knowledge it will reveal, not because a sellable product is within the researcher's grasp.

This was Vannevar Bush's point in *Science, The Endless Frontier,* a report he produced in 1945 for President Truman. Bush, director of the Office of Scientific Research and Development, laid out a vision for the nation's scientific future in which government-financed basic research produced (eventually) the fuel for a vigorous economy.

To cut a long and complex story very short, the result of all this cogitating was the creation of the National Science Foundation, a government-financed organization that, since its establishment by Congress in 1950, has been the major source of federal funding in the physical sciences.

The National Institutes of Health, which finds its roots in nineteenth-century federal health agencies, achieved something like its present form in the years after World War II, when Congress established institutes dealing with a variety of medical conditions.

For decades, through these and other agencies—like the Defense Advanced Research Projects Agency (DARPA), the Office of Naval Research, the U.S. Geological Survey, the National Oceanic and Atmospheric Administration, among others—the federal government had the preeminent role in financing fundamental science and a big role in financing research in universities and labs. And the United States led the world in both basic scientific research and the creation of new technologies derived from it.

Although some challenge the idea, there is wide agreement that federal support of basic research was driving prosperity as well. The Bipartisan Policy Center, a nonprofit that seeks to identify practical steps to promote health, security, and economic opportunity, divides current federal research spending into four categories:

- Military. About half of federal research dollars go to the Department of Defense. Much of the work involves technology to improve national security, the internet, the Global Positioning System (GPS), and many other technologies in wide use in civilian life. These and others originated in the Defense Advanced Research Projects Agency, which functions as a kind of Pentagon skunk works. (It is impossible to overestimate the importance of DARPA, where, among other things, computers first learned to send each other messages.)

- Health. Through the National Institutes of Health, the federal government is a major source of funding for health-related research.

- Space. The National Aeronautics and Space Administration (NASA) is best known for space flight, but for decades it has also been on a "mission to Planet Earth" to study the environment, particularly the atmosphere in an era of climate change. Needless to say, this research has attracted more than a fair share of conservative congressional wrath.

- Basic. In many ways, this is the most important funding category, because it includes work that has few obvious immediate applications and is therefore not a favorite of private industry. Much of the money for basic research is funneled through the National Science Foundation, whose elaborate review process for grant applications is widely (if sometimes reluctantly) regarded as a model.

People argue about whether the federal research budget is adequate, and whether it is being pinched. Advocates of more spending will compare today's levels of federal research spending against its high water mark in the 1960s, when the United States spent about 2 percent of GDP on research. They normally neglect to add that we were in a space race then, a kind of artificial goad for research spending. Today, according to the policy center, the figure is under 1 percent.[5] Otherwise, the research share of federal discretionary spending (money left over after nonnegotiable items like Social Security and interest on the national debt are accounted for) has remained pretty stable.

Generally, academic researchers use some of this federal largesse to undertake work that was unlikely to be financed by the private sector. In the years after World War II, private companies like AT&T and General Electric ran their own in-house research labs producing all kinds of useful inventions and knowledge, much of which did not relate to their own products. My favorite example of this phenomenon occurred in 1964 at an AT&T Bell Labs facility in Holmdel, New Jersey, where efforts to rid a dish antenna of an irritating "hum" resulted in the discovery of microwave background radiation, the cosmic echo of the Big Bang. The discovery did nothing for the telecommunications industry, but in 1978 the AT&T scientists won the Nobel Prize.

This landscape began to alter around the end of the Cold War, when private money began to hold more and more sway in American research.

One important factor was legislation to encourage technology transfer, in particular the Bayh-Dole Act of 1980. The legislation, sponsored by Senators Birch Bayh, an Indiana Democrat, and Bob Dole, a Kansas Republican, was a response to complaints that the fruits of federally financed research were not finding their way into the private sector, to industries that might make profitable use of them. In part, that was because of government restrictions on the licensing of technologies created with federal funds. Except in relatively rare circumstances, the government would not readily transfer ownership of federally funded inventions, even to the inventing organizations. Instead, they issued nonexclusive licenses.

But companies did not want to invest time and effort in products relying on technology over which they had little control. As a result, much of the technology developed with federal research funds languished. According to the Council on Governmental Relations, an organization of research universities, in 1980 the federal government held title to approximately 28,000 patents.[6] Fewer than 5 percent of them were licensed to industry.

Under Bayh-Dole, researchers supported with federal funds can move to establish ownership interests in the products of their work and, therefore, bend it to commercial applications. It has produced chemotherapy agents, vaccines, recombinant DNA technology, and a host of other applications. Thousands of companies have been formed to capitalize on these developments, resulting in the creation of tens of thousands of jobs.

"The civilian research agencies were designed as temples of scientific excellence and technological prowess, but they . . . are ill-structured to create and sustain essential links between knowledge generation, technological innovation and desired social outcomes," Daniel Sarewitz, co-director of the Consortium for Science, Policy, and Outcomes at Arizona State University, wrote in a column in the journal *Nature*.[7]

As the Council on Government Relations put it, "(c)ertainty of title to inventions" made with federal support protects scientists who continue to work with the technology and build on it. "Accordingly the Bayh-Dole Act continues to be a national success story . . ."[8]

There is nothing inherently wrong with private spending on research. The problems occur when funders tie strings to the researchers whose work they finance. They may require researchers to keep their results secret, on the grounds that the information is proprietary—a practice that would until recently have been antithetical to research.

Also, even high-minded researchers can become contaminated with conflicts of interest, whether they realize it or not. As the authors of a paper in the *New England Journal of Medicine* put it in an article on the use of certain drugs in the treatment of cardiovascular disorders, there is "a strong association between authors' published positions on the safety of calcium-channel antagonists and their financial relationships with pharmaceutical manufacturers."[9]

Conflict of interest now so pervades science that medical journals, universities, and even government agencies like the Food and Drug Administration no longer even attempt to ban it. They may insist on disclosure, but that requirement can be difficult or even impossible to enforce. Conflict of interest arising from financial ties is "widespread," the *Journal of the American Medical Association* reported.[10] Other studies have reported that gifts as modest as notepads or ballpoint pens seem to influence prescribing habits of physicians.

A corollary of Merton's norm of communalism is openness. "The communism of the scientific ethos is abstractly incompatible with the definition of technology as 'private property' in a capitalistic society," he wrote.[11] As a result, findings are shared openly.

But according to Sheldon Krimsky, a social scientist at Tufts University who has studied the issue, financial considerations have greatly modified the way researchers do their work and even the way they discuss it with their colleagues and research peers. "Scientists at one leading research university reported that their colleagues were loath to ask questions and give suggestions in seminars or across the bench," he wrote, "for there was a feeling that someone might make money from someone else."[12] There are even tales of students working in high-performing university labs who run into trouble writing their own course papers and theses because of nondisclosure agreements.

Private money can influence the conduct of research in other unfortunate ways. For example, an embarrassing uproar broke out at the Oregon State University College of Forestry when researchers there studied the effects of what is called "salvage logging," the removal of timber after a major fire. The test case for Dan Donato and his colleagues was the Biscuit Fire of 2002.[13]

The logging industry likes salvage logging—apart from the wood it produces it has the added benefit (proponents say) of encouraging forest regrowth and reducing future fire risk. But the researchers found that heavy equipment used to remove the fire-damaged trees killed seedlings and left debris that actually increased the fire hazard.

Other researchers at OSU said the paper was flawed and asked the journal to delay publishing it in print form, a request, *Science* said, "widely perceived as an attempt at censorship." (A tax on logging business activity provides 12 percent of the forestry college's budget.)

Before the dust settled and the paper was published in 2006, the U.S. House Committee on Natural Resources held hearings in Oregon (October 24) in which Donato was subjected to intense questioning. Memos critical of his work, some "quite personal," were posted anonymously, and the forestry school's dean found himself attempting to play down memos referring to environmental activists as "goons."[14]

After the paper was published, the federal Bureau of Land Management pulled an Oregon State University (OSU) grant, saying it had somehow violated the agency's agreement with the university. The Union of Concerned Scientists, which examined the affair as a case study in its work on research integrity, offered a different explanation: Donato's conclusion "clashed with the Bush administration's support for easing restrictions on such salvage logging."[15]

Or consider the Weed Society of America, a research organization. Its members have often noted that big agricultural companies are a major source of money for research on weeds. But if your money comes from Big Ag, you are probably going to be researching chemical herbicides rather than natural methods of weed control.

As Krimsky put it, "if a subfield of science is largely supported by industry, then it scarcely matters whether scientists are working in corporate facilities or university facilities. The problems framed, the type of data collected, the effects measured are largely under the influence of the industry sponsor."[16]

Perhaps a more important issue is a widespread tendency for people to unconsciously produce research results that will be welcome to their sponsors. In "Conflict of Interest and the Unconscious Intrusion of Bias," Don A. Moore and colleagues said the researchers they studied did not knowingly subvert their findings in the service of their funders. (What they actually said was, "conflicts of interest did not induce the hypocrisy of public misrepresentation in the presence of private honesty.") Instead, they wrote, the conflict "induced bias in both public and private judgments."[17]

Moore and his colleagues are hardly the only researchers to come to this conclusion. Lenard I. Lesser and colleagues studied research on nutrition—interventional and observational studies of soft drinks, juice, and milk—as well as scientific reviews of such work. Some were financed by food producers, some were not. Among other things, Dr. Lesser and his colleagues looked at the relationship between the funding source for the studies and their conclusions. Among studies carried out without food industry funding, 37 percent came to what they called "an unfavorable conclusion." Among those with industry funding, the unfavorable percentage was 0.[18]

In his book, Krimsky reported that by the 1980s 31 percent of MIT's biology department had ties to private companies (twenty-seven of them), and that at Stanford, about twenty faculty members were involved with twenty-five firms. At Harvard, the figures were 20 percent, at forty-three firms.

In an extreme case, in 1998 the pharmaceutical firm Novartis made a deal with the University of California–Berkeley pledging to give the university $25 million over five years in exchange for "significant access" to the work of the school's microbiology department, its laboratories, and discoveries.[19] The idea was greeted with outrage at the time, but other researchers who studied it later said it had produced few benefits for Novartis (or its successor, Syngenta). The university benefitted from higher stipends to graduate students.

Still, the report, by Lawrence Busch of Michigan State University and colleagues, recommended that Berkeley avoid future industry agreements "that involve complete academic units or large groups of researchers."[20]

When Vannevar Bush advocated decades ago for greater federal support of basic research, he also urged substantial "overhead" payments to universities to cover the costs of maintaining institutional infrastructure. In practice, faculty members doing research in federally fundable science and engineering fields are sources for these overhead funds. Professors in departments of literature, history, and art don't have this financial potential. And that can have unintended consequences.

Derek Bok, then the president of Harvard, raised the issue in a report to the university's Overseers—its trustees—in 1981. He said he had "an uneasy sense that programs to exploit technological development are likely to confuse the university's central commitment to the pursuit of knowledge and learning by introducing into the very heart of the academic enterprise a new and powerful motive, the search for utility and commercial gain."

He was prescient but also, I fear, naive.

"The real story in academia today is the story of the systemic transformation of our universities into market-oriented enterprises with multibillion dollar budgets and vast investments in such emerging fields as biotechnology," Andrew Delbanco wrote in the *New York Review of Books*. In the new university, he wrote, all too often "the humanities are not really part of this story except in the sense that they are being marginalized by it."[21]

In this new university paradigm, teaching is often a subsidiary activity conducted much of the time by part-time "adjuncts" who may receive low pay and few, if any, benefits and who may have little connection to the life of the university or its students, possibly because they are working two or more jobs. (The rise of online instruction may ultimately put even these ties to the test.)

The influence of private money in research can be particularly pernicious when it comes to medicine. For example, at one time the Food and Drug Administration sought to staff its expert advisory panels with researchers with no financial conflict of interest. Over the years, as it became more and more difficult to find knowledgeable people without such conflicts, the agency has relaxed its rules. First, a financial interest was okay if it was "remote," then it became accepted if the need for the particular person outweighed the conflict, and finally the agency just required people to disclose their interests. This greater transparency seems to be the favored remedy for virtually every problem. But as I know from experience, not everyone obeys.

For us at the *Times,* reporting on these conflicts has been surprisingly complex.

First, how would we know? Many journals require researchers to disclose financial arrangements they have with companies with an interest in the research and—I guess—they usually do. But we know there's no guarantee. So, how much reporting time should we devote to ferreting out the financial interests of the researchers we are writing about? Most science journalists do not have a lot of extra time.

Also, once we have the information, how much space should we devote to it? This problem is a killer in print media but it does not go away when you move online. If you only have 600 words to describe some bit of research, you may be unwilling to spend even a hundred of them on financial matters.

And, just because a researcher does not now have a tie to a given company, that does not mean she is not hoping to acquire a research grant from the firm—and is altering her work to suit.

Finally—and for me this is the most important issue—saying that a researcher was financed by someone with an interest in the outcome of the research tends to put the work under a cloud. People will assume there must be something wrong with it, even if it was conducted according to the rules of the game and its results were presented fairly.

This issue is one with no easy answers.

By now, the National Science Foundation can fund only about a fifth of the research proposals it receives, even though most are eminently worthy of support. Reviewers have begun to turn away from young researchers in favor of older scientists with safe track records, even though high-risk work, often by young researchers, is disproportionately the source of game-changing findings.

This situation should not be surprising. In an era of stagnant funding grant-makers may be wise to aim for recipients who are more or less certain to produce publishable results of some sort, even if they are incremental findings that don't really move the metaphorical ball very far. The median age of first-time recipients of an NSF grant has been rising and rising—it is now in the mid-40s.

Also, growing inequality means "more successful scientists are much more likely than less-successful ones to be centrally located in global collaborative networks," as Yu Xie, a sociologist and statistician at the University of Michigan, wrote in the journal *Science*.[22] So even though international collaborations are more possible these days than ever, especially with the growth of the internet, the ability to benefit from these new connections varies according to the place a researcher occupies in the global research hierarchy.

In recent years people have proposed a number of steps they say would reduce the pernicious effects of money on academic research, including making management of the grant-making process more transparent, new conflict-of-interest regulations, more federal oversight, and changes in the Bayh-Dole Act to discourage what some call reflexive patenting. Some have even suggested a mechanism by which the National Institutes of Health could intervene if it thought patent restrictions were standing in the way of potentially useful research.

A few years ago, when I was teaching at Harvard, a faculty friend told me that Larry Summers, then the university president, had asked the university's professors to let him know what they were working on that might produce money. My friend said he replied: "I am happy to say, nothing."

Selling Health

Q: What is the definition of a healthy person?
A: Someone who has been inadequately worked up.

I thought of this old medical joke when a friend of mine developed excruciating back pain. Magnetic resonance imaging (MRI) disclosed that one of the disks in her spine was damaged. She was admitted to a major medical center and prepared for back surgery. But because she was running a slight fever, she could not have the surgery that night. Instead, she was given painkillers and sent home.

That's where I visited her the next day with another friend, a physician whose specialty is pain. He concluded that her problem was muscle spasms in her lower back and not a bad disk. He treated her for the spasms—the treatment was simple and did not involve drugs—and she immediately felt much better. (She never did have the surgery, and today she is fine.)

As my doctor friend and I walked back to my home, I wondered to myself how a muscle-related treatment could do so much good for someone whose disk damage had been documented by an MRI. Suddenly the answer dawned on me. I asked my friend, "how many people are walking around in fine form, even though they have disk damage?"

After age 50, about 35 percent, he said. So I asked him, "how many people should have disk surgery?" He replied, "practically nobody." Researchers in Japan who studied seemingly healthy, pain-free people found disk degeneration in 17 percent of men and 12 percent of women *in their twenties.* By age 60, the numbers had risen to 86 and 89 percent.[23]

Though underlying rates of disk damage do not appear to vary much from country to country, rates of spine surgery are highest in the United States. According to the National Institutes of Health, "There remains little or no medical, clinical, or surgical evidence to support such variability."[24]

A few years after my friend's experience, a sports orthopedist in Gulf Breeze, Florida, reported what happened when he did MRI scans on the shoulders of thirty-one professional baseball pitchers, none of whom was having pain or any other shoulder problem. The images showed abnormal shoulder cartilage in 90 percent, and abnormal rotator cuff tendons in 87 percent. "If you want an excuse to operate on a pitcher's throwing shoulder, just get an MRI," the doctor said.[25]

Ordinarily, true double-blind clinical trials of surgical procedures are not done, because they would require patients in the control group to undergo pointless "sham" surgery with the attendant risks of anesthesia and a hospital stay. But in a report in 2002 researchers in Texas described what happened when they subjected people with arthritic knee pain to a randomized clinical trial in which some underwent one of two forms of arthroscopic knee surgery and the third group underwent a sham operation. At several points over the next two years, patients and medical practitioners assessed the results in terms of pain, function, and tests like climbing stairs. "At no point did either of the intervention groups report less pain or better function than the placebo group," the researchers reported.[26]

In 2008, another study found that people treated for arthritis pain in the knee did just as well with physical therapy and anti-inflammatory medication as they did with surgery.

The story of medicine is replete with the enthusiastic embrace of tests or treatments that are oversold, worthless, or even harmful.

There was a brief mania a few years ago for spiral computerized tomography (CT) scans to detect lung cancer. It was introduced around 2000 to enable doctors to spot tiny lesions in the lungs that had gone undetected in X-rays and other tests. The doctors hoped the new test would help them detect lung cancer while it was still small enough to be removed. Advocates said the procedure could cut deaths from lung cancer by 80 or 90 percent. People lined up for the scans.

Initially, for Barnett Kramer of the National Cancer Institute, it was far from clear that the test would turn out to be worthwhile. Rather, it was another example of how medical procedures can come into wide use before there is any evidence that they actually do any good for anyone other than the people selling them. As he noted later, "there are few issues in medicine that polarize the public, their elected representatives and health professionals more than medical screening."[27]

Kramer, a physician with a master's degree in public health, is, among other things, director of the institute's Division of Cancer Prevention. He has run consensus conferences at which experts debate the best tests and treatments for malignancies. But when he spoke about cancer screening at the MIT conference on medical evidence he emphasized that he was speaking for himself. Many doctors do not like to hear what he has to say about cancer screening.

Tests can detect abnormalities, he said, but they do not necessarily differentiate between trivial changes and signs of a true problem, something "clinically significant." With the spiral CT, he said, "we are picking up tiny lesions but we don't know what their natural history is, because we have never been able to watch them before." Kramer also noted that with some exceptions (brain cancer and leukemia are two), the primary cancer itself is not what kills people. The cancer's spread, its metastases, do the deadly work. If a cancer has already seeded metastases in the body, he said, "picking up a cancer early is not equivalent to picking up a cancer early enough." And if the cancer was never going to spread, picking it up doesn't help the patient either. The doctor has simply identified something that was never going to be a problem.

"Well," one might say, "so what?" After all, it is better to be safe than sorry. But being safe carries a cost, in this case the cost of screening lots of people for every one who turns up with an alarming symptom. Often the tests lead to further tests or treatment. And cancer treatment is not just expensive, it is risky. In the case of lung cancer, it may involve surgery, which carries its own risks, and exposure to toxic chemotherapy drugs and radiation. Along with radiation and chemotherapy, breast cancer treatment may involve removal of one or both breasts. Prostate cancer treatment leaves many men incontinent, impotent, or both.

When the CT-scan-for-lung-cancer boom began, marketers of the tests sold them enthusiastically, and plenty of people had them. Some were found to have lung lesions and underwent surgery to remove them. They probably believe the surgery saved their lives. But it is impossible to tell.

In 2010, the National Cancer Institute released initial results from a large study that concluded that yearly low-dose helical computed tomography could reduce lung cancer mortality by 20 percent among heavy smokers.[28]

"It is clear" Kramer and his co-authors wrote in 2011, that CT scans to detect lung cancer were not as effective as people had hoped, but not as worthless as some had feared. They added, "such is the power of a randomized controlled trial." They noted, though, that issues like cost-effectiveness and the adverse effects of treating overdiagnosed cancers, among others, remain to be discussed. In addition, the National Cancer Institute pointed out the obvious: "the single best way to prevent lung cancer deaths is to never start smoking, and if already smoking to quit permanently."[29]

The argument over two more widely used tests—mammography for breast cancer and the PSA antigen test for prostate cancer—continues. Medical organizations of all kinds have been banging the drum for years for both of them, on the grounds that early detection greatly increases the chance of cure. And people whose cancers are found early do survive longer than those whose are not. It seems to make sense. But does it?

Screening will always advance the date of diagnosis. So if you measure survival from the time of diagnosis you will always see a benefit for screening. But the patients are not necessarily living any longer than they would have anyway. They are just finding out sooner that they have cancer.

Also, even among people with the same cancer diagnosis, some have cancer that is fast and deadly while others have indolent tumors. Screening differentially turns up indolent cancers. But for many cancers, like breast and prostate cancer, we don't yet know how to tell which is which.

"You need a randomized, controlled clinical trial," Kramer told us at the boot camp on medical evidence. "Start everybody at the same time and you count dead bodies." He might have been right, but it would probably be impossible to attract enough people to participate in a trial in which they would agree to be randomly screened or not screened.

The prostate-specific antigen test for prostate cancer came into wide use in the 1990s, when the Food and Drug Administration authorized the routine use of a test for blood levels of prostate-specific antigen in men 50 and older. The number of prostate cancer diagnoses rose rapidly. Thousands of men received treatment far earlier than they would have otherwise. But the prostate cancer death rate, in aggregate, barely budged.

At the MIT conference, Kramer told us he had decided years ago he would not take the PSA test. But it was a decade or more before the medical establishment came to embrace his view, or something like it. Finally, in 2011, the U.S. Preventive Services Task Force, a volunteer panel of experts, said that healthy men should not be screened because, overall, mass screening would do more harm than good. The American Urological Association reacted with outrage, but eventually said that screening should be recommended primarily for men between 55 and 69 and that the test should be done every two years, not annually.

Similar issues arise with screening mammography—that is, mass routine screening for women who do not have any symptoms of breast cancer. There is evidence that yearly screenings do save lives among women aged 50–64. But if they do, the effect is relatively modest, at least in aggregate. (Granted, if you believe you are one of the ones who benefit, the gain does not seem modest at all.) Breast cancer mortality has fallen from about 32 per 100,000 people in the mid-1970s to about 20 per 100,000 people by 2010 according to the National Cancer Institute. But it is hard to say how much of that decline is owing to earlier diagnosis and how much to better treatments.[30]

The mammography issue came home to me when I was invited to take part in a panel discussion on the subject, to be held at the Mayo Clinic, the eminent hospital and research center in Rochester, Minnesota. The panel had been organized to attack coverage of mammography, particularly ours, particularly related to research on the effectiveness of mammography in mass screenings. The work, reported in the *Lancet,* a British medical journal, was by the Nordic Cochrane Collaborative, the Denmark-based arm of a respected international organization that studies research reports to determine whether their data support the claims made for them.

The group's mammography findings were shocking. The collaborative group had looked at leading studies of mammography and concluded their methods were such that the studies could not be said to show what screening advocates said they did. The report concluded, "there is no reliable evidence that screening for breast cancer reduces mortality."[31] The collaborative noted that advocates of mass screening say detecting small cancers early reduces the need for mastectomies. The Cochrane group said mastectomy rates rise in areas where mass screening is routine.

The group took another look at the issue again in 2006 and declared their new review "confirmed and strengthened our previous findings." Finally, ten years after the original report, they looked again. In an essay summarizing this review's findings, Peter C. Gøtzsche, director of the Nordic Cochrane Centre, said there did not appear to be a decrease in advanced cancers, something that would be expected in areas with mass screenings. They attributed declines in death rates to improvements in treatment, particularly chemotherapy agents.

"Therefore," he wrote, "what was considered so controversial in 2001 is now increasingly being recognised to be true, even by people who advocated the introduction of screening in the first place." He added, "it is now essential that women be provided with information that allows them to make an informed choice about mammographic screening, rather than being pushed toward mammography as routine, while being told it is an unambiguously beneficial test."[32]

In 2015 researchers reported the results of a study of twenty years worth of data on 100,000 women whose mammograms showed they had a condition called ductal carcinoma in situ. Most American women who receive this diagnosis have some kind of treatment, often lumpectomy and radiation, sometimes mastectomy. Some adopt a watch-and-wait approach, a stance becoming more common as more and more doctors reject the idea that DCIS is cancer at all.

The new research, reported in *JAMA Oncology,* found that the death rate from breast cancer among these women was 3.3 percent—just about identical to the breast cancer death rate in the population at large.

As I write, women seeking guidance on mammograms will learn that they should have them annually starting at 40 (American College of Obstetricians and Gynecologists), annually starting at 45 and every two years after 55 (American Cancer Society), or every other year starting at age 50 and ending at age 74 (U.S. Preventive Services Task Force). It's no wonder women are confused.

Something similar played out with the use of bone marrow transplants as a treatment for breast cancer. In this treatment, which

first came into use in the 1980s, women are subjected to chemotherapy strong enough to kill all the cancer cells in their bodies.
Such strong treatments also destroy their bone marrow, which must
then be replaced in a transplant. When the treatment first came
into use, without clinical trial evidence that it would be helpful, insurance companies refused to cover it. Under an onslaught from
the breast cancer lobby, they relented. Years later, researchers determined that bone marrow transplants offered no survival advantage
and the practice has fallen out of favor.

The breast cancer lobby is an excellent example of a politically
powerful patient advocacy group driving government science
spending. In the early days of the AIDS epidemic, when deaths
were rising dramatically and doctors had little to offer desperate patients, AIDS activists modeled their lobbying efforts on breast
cancer campaigns, which did a lot of good.

The breast cancer lobby has been so effective that today, many
Americans believe breast cancer is the leading cause of death
for American women. It isn't. It isn't even the leading cause of
cancer death among American women—that honor goes to lung
cancer. The leading cause of death for American women is heart
disease. About half of American women die of it. But heart disease in women—not necessarily the same, by the way, as heart
disease in men—receives far less attention.

By now, many people who were once enthusiastic advocates of
cancer testing concede that the practice may have been oversold.
Here as elsewhere in medicine, the issue is surrogate end points.

"Early detection's mission is to reduce mortality; we are failing
at that," Kramer says. If the goal is to reduce cancer morbidity—the
ill effects of living with the disease—tests that end up subjecting
people to arduous treatment for something that was never going to
hurt them have the opposite effect. The PSA test is particularly sus-
pect, given the prevalence of prostate cancer, especially in older
men, and the potentially devastating side effects of its treatment.

The advent of the CT scan has resulted in an increase in kidney cancer diagnosis, because the tumors turn up while doctors are looking at or for something else. Death rates for the disease are more or less unchanged. The explanation seems to be that many kidney cancers are indolent, nonthreatening. Doctors call them "incidentalomas." But once people know they have a malignant tumor in one of their kidneys, they want it treated, usually by removal of all or part of the affected organ.

Cancer screening is far from the only kind of medical procedure sold to the public before its benefits, if any, have been demonstrated with end points we care about. Overselling medicine has a history about as long as medicine itself.

Consider what happened with the invention of X-rays in the nineteenth century. Until then, physicians trained by dissecting cadavers. X-rays gave them a different view of our insides and some of them found it alarming. Instead of being neatly arranged, like the organs of a cadaver lying on a dissecting table, organs of living people sometimes looked like they were "sloshing around," as Mary Roach puts it in her excellent tour of the digestive tract, *Gulp.*[33]

What was going on? The answer turned out to be simple. People stand for X-rays and their organs move a bit under the influence of gravity. But doctors began performing surgery to correct "dropped organs," a nonexistent condition.

Or consider the famous inkblot or Rorschach test, devised almost a century ago by Hermann Rorschach, a Swiss psychologist whose patients included people with schizophrenia. The test is one of several commonly used "projective" tests, so called because they ask people to look at an image whose meaning is not obvious and interpret or "project" meaning onto it. In the Rorschach test the image is an inkblot—ten particular blots have been used from the outset. In another projective test, the Thematic Apperception Test, people must interpret pictures depicting domestic scenes, telling a story about each.

Advocates of the Rorschach test say it enables clinicians to gain information about patients' psyches unobtainable from interviews or tests in which patients report their feelings. The test is particularly useful, its advocates say, with people who have something to hide—criminal defendants, say, or even some parents in custody disputes. Rorschach advocates concede that a lot depends on the insight and skill of the clinicians who administer the test, but they note that the same can be said about other kinds of mental health assessments.

Over the years, though, the test has been taken apart by reviewers who question its usefulness. Until standardized scoring rules came into use in the 1970s, it was difficult to compare results when the test was administered by different clinicians. The standards made that chore easier, but now, of course, they have turned up online, allowing people about to be tested to concoct responses most likely to generate the score they want.

In a review of research on the test, which was reported in 2001 in the journal *Psychological Science in the Public Interest,* Scott O. Lilienfeld, a psychologist at Emory University, and his colleagues concluded that data on the test and its findings do not support its widespread use.[34] "They are not as useful for most purposes as many clinicians believe," Dr. Lilienfeld said in an interview later.[35]

The researchers cited one study of 123 people who were given the test, most at a California blood bank. Though these subjects had no history of mental illness or other psychiatric issues, many of them were scored as having clinical depression or signs of extreme narcissism. Sixteen percent scored as "abnormal" in the test's schizophrenia index. (The rate of schizophrenia in the general U.S. population is about 1 percent.)

Other researchers say that under the standards enunciated in the *Daubert* case, the test should not be admissible in court.[36] But it is still in use, and its results are taken seriously in criminal courts, family courts, and elsewhere.

Evidence-Based Medicine

When I first heard of "evidence-based medicine" I was shocked—shocked, that is, to discover that there was any other kind. Shouldn't all medicine be based on evidence? Well, it isn't. Much of what we think of as medical science is really medical lore.

"Right now, health care is more evidence-free than you might think," Peter Orszag, who had directed the Office of Management and Budget, wrote on the op-ed page of the *New York Times* shortly after leaving the White House.[37]

His assessment applies far beyond the supplement industry which, through the influence of its powerful allies in Congress, manages to avoid the kind of regulation the pharmaceutical industry complains about. Nor is it limited to so-called "alternative treatments" like aromatherapy or chiropractic or acupuncture, although in 1993, when the NIH formally established the National Center for Complementary and Alternative Medicine, critics declared it was a waste of scarce research funds. (From time to time the institute reports that this remedy or that is worthless. For example, center-sponsored tests found that echinacea does not cure the common cold. On the other hand, center-sponsored studies have shown that some alternate treatments, notably acupuncture, are quite effective.)

But, as Orszag wrote in his op-ed, "even when evidence-based clinical guidelines exist, research suggests that doctors follow them only about half the time."

Why is this? For one thing, many practitioners rely on what they learned in medical school or from respected mentors in their training. They may be unaware of information in practice guidelines, which can take years to make their way from the councils of professional societies or federal advisory panels to the bedside.

A prime example is the annual physical. Except as a better-than-nothing way of building a relationship with a physician, the annual physical accomplishes little, health experts have asserted for years. There is little evidence that the routine tests in these exams extend survival, and even if the tests are useful they don't have to be done every year. Yet, annual checkups for people with nothing wrong with them are the most common doctor visits in the United States.

Something similar is true with the Pap test to detect cervical cancer. The test is a genuine lifesaver, and researchers are working now to devise low-cost versions that can be used in poor countries, where cervical cancer is a leading cause of death among women. But in developed countries, women who have had normal results don't need the test every year. And they don't need them at all if they are not sexually active or are in a trusted monogamous relationship. But it's hard to break these medical habits. In the United States, even women whose cervixes have been removed find themselves treated to a Pap test. Or, as a letter-writer put it in the *Times*'s letters column, "it is hard to change profitable behavior even though new studies prove it to be unnecessary or harmful."[38]

The Affordable Care Act passed in 2009 establishes the Patient-Centered Outcomes Research Institute, tasked with coordinating research on the effectiveness of various medical treatments. But, as the journal *Nature* noted, "comparative-effectiveness research has proved controversial." In theory, it would determine whether one treatment is better than another and cut spending on treatments whose value is limited or even nonexistent. But people eager to criticize the Obama administration seized upon this provision of the legislation, saying it would produce what amounts to rationing of health care.

That is too bad. Public health officials need evidence for alternative strategies, data to help them allocate scarce resources, and information to share with key stakeholders.

THE PLACEBO EFFECT

In many clinical trials of drugs or other treatments, researchers compare a test group of patients receiving the intervention to a control group receiving a sham treatment or placebo. In theory, people receiving a useful treatment will do better than those receiving a dummy pill. Unfortunately for this theory, people given a sham treatment may feel better as well, a phenomenon called the placebo effect.

The placebo situation is not simple. Some people suggest that there may be no such thing as a placebo effect. We see it, they say, only because few medical conditions take an unremittingly down-ward course. Many people, even in the grip of illnesses that will eventually kill them, have good days and bad days. If you have such a disease and receive a placebo and then have some good days, it may look like you have benefitted from a "placebo effect."

Others say placebos can be so effective doctors should offer them as treatments without informing their patients, especially if the pa-tients have conditions for which no good treatments exist. That was what the German Medical Association recommended in 2011.[39]

This idea has been greeted with skepticism. "That's what I call lying," said Ted Kaptchuk, a professor at Harvard Medical School and director of the university's program in placebo studies. "It would be unacceptable in the United States."[40]

Then there is the so-called nocebo effect—the idea that when you know you are taking a drug, you will feel side effects, even if you are receiving a dummy pill. That is why many people receiving placebos in clinical trials drop out. They say their side effects are too difficult to bear.

It would be unfair to describe all or even most dietary supplements as snake oil. At least I hope it would be. I hope the people selling them actually believe in them. But in my opinion it is not a coinci-dence that the supplement industry lobbied so effectively to escape rigorous regulation by the Food and Drug Administration. Its de-cisive victory came in 1994 with the Dietary Supplement Health and Education Act. Under the act, supplements are classified as food, not drugs, and they are virtually unregulated. As long as makers don't claim they treat disease, they don't have to prove any-thing about their products.

When President Clinton signed the measure, he praised it as bringing common sense to the industry, but the act has been widely criticized. Absent regulation, few can say with confidence what these products will do for them, or even whether the pill in the bottle offers the product in the dose the label specifies. But never mind all that. The supplement industry has sold itself so assiduously that by some estimates about half of American adults take supplements of some kind.

Many people regard dietary supplements as more or less benign. Today we know that some vitamins, vital for good health, can cause problems if taken in large quantities. The effect is particularly noticeable in fat-soluble vitamins like A and D and E. Others, like Vitamin C, are water soluble. If you take more Vitamin C than you need, you will eliminate the excess in your urine. Your body will store A, D, and E in your fat or adipose tissue—sometimes to levels that can be unhealthy.

Because the supplement industry is more or less unregulated, it was not a surprise when in 2015, a study reported in the *New England Journal of Medicine* found that more than 20,000 Americans end up in the emergency room each year suffering the ill effects of supplements.[41] About 2,000 of them end up hospitalized.[42]

Still worse are certain weight-loss supplements, which can cause chest pain, heart palpitations, and irregular heartbeats. These products are taken most often by young people, some of them high school athletes. These students are lucrative targets for companies that market so-called sports nutrition supplements to help young athletes recover from strenuous workouts, build muscle, or improve endurance. These products are particularly concerning because there is virtually no research on how any of these supplements affect growing bodies.

For homeopathy, as with supplements, I hope the people endorsing it actually believe in it, discredited as its founding principles may be. Homeopathy is based on two ideas: that "like cures like"—that tiny quantities of a substance which gives healthy people a disease will cure the disease in those who fall ill with it; and the so-called "law of minimum dose," which holds that the more a treatment is diluted, the more effective it becomes. Adherence to this "law" often results in homeopathic remedies so diluted that, as the National Institutes of Health puts it, "no molecules of the original substance remain."[43]

The roster of people who adhere to this obviously faulty concept includes some surprises: Prince Charles is an advocate, for example, and Luc Montagnier, the virologist who won the Nobel Prize for identifying the virus that causes AIDS, has written approvingly of it. Regardless, as the NIH says, it is based on ideas "inconsistent with fundamental concepts of chemistry and physics."

The fact that a homeopathic remedy is ineffective does not necessarily mean it is harmless. The NIH warns that "some products labeled as homeopathic can contain substantial amounts of active ingredients and therefore could cause side effects and drug interactions."[44]

There are plenty of other snake-oil remedies—copper bracelets for arthritis, magnets for pain relief, and so on. My favorite snake-oil therapy is a kind of "therapeutic touch" in which the practitioner waves his or her arms or hands in the air above the prone body of the sufferer to manipulate the person's supposed energy field. This loony treatment was famously debunked by a 9-year-old who, with the help of her mother, a registered nurse, subjected the technique to a clinical trial—part of her fourth-grade science project.[45]

People who peddle this kind of thing can always explain away any outcome of its use. If the patient's condition improves, the treatment worked. If the patient stays the same, the treatment stabilized the condition. If the patient deteriorates, the dose or application was wrong. And if the patient dies, the magic elixir was unfortunately applied too late.

Every now and then the Federal Trade Commission or some other watchdog agency calls out the maker of a snake-oil remedy. Meanwhile, you can protect yourself by not taking any remedy without your doctor's knowledge. Even seemingly benign remedies like vitamin supplements can cause unwelcome interactions with other drugs.

Be especially suspicious if you learn about a remedy from a commercial on late-night television or the like, or even from a seemingly reliable news organization, rather than from a health care provider you trust. Researchers who studied newspaper and television coverage of drugs for the prevention of cardiovascular disease and the prevention and treatment of osteoporosis found that some news organizations tended to emphasize benefits over potential harms. In half of the stories they analyzed, someone cited as a source turned out to have financial ties to the drug's maker.[46]

Be suspicious, generally, of any prescription medicine advertised directly to consumers. Until 1985, this practice was illegal in the United States, and even today it is allowed in only one other country, New Zealand. Advocates of the practice, many of whom are connected in one way or another to the pharmaceutical industry, defend it as offering consumers yet another source of information about drugs they might find useful. Critics say judgment on these matters is best left to health care professionals.

According to the World Health Organization, direct-to-consumer advertising of prescription drugs does not inform patient choice so much as drive it—"typically in the direction of expensive brand-name drugs."[47] Plus, the advertising "overshadows the fact that less-expensive drugs and nondrug therapies can often do just as well."[48] The WHO said surveys in New Zealand and the United States show that if patients ask for a drug by name it is "very likely" their physicians will prescribe it for them. In fact, practitioners who find themselves pestered by patients demanding a drug may prescribe it, even if they believe it is unsuitable.

The U.S. Congress investigated the issue in 2008 after the pharma giant Pfizer ran advertisements promoting the cholesterol-lowering drug Lipitor and featuring Dr. Robert Jarvik, identified as the "inventor" of the artificial heart, even though, as the WHO put it, he had "never been licensed as a medical doctor, could not legally prescribe anything and was not the inventor of the artificial heart (at least according to three former colleagues . . .)." Pfizer pulled the ad.

But as anyone who watches commercial television in the United States knows only too well, the practice remains widespread. Recently, for example, there has been a lot of hype—and a lot of advertising—about drugs to treat a condition known on television ads as "low T"—low levels of testosterone. There is such a condition; men who develop it can become infertile. But relatively slight declines in testosterone also occur naturally with age. In an example of what some have called "disease-mongering," this decline is now described as a pathological condition treatable with a prescription for testosterone. It is administered in patches, gels, and in injectable form to the tune of billions of dollars each year in the United States.

What the ads neglect to point out is this treatment can significantly raise heart attack risk. As one doctor put it on the op-ed page of the *New York Times,* testosterone is "a risky drug to treat a trumped-up disease."[49]

In 2015, the American Medical Association called for a ban on direct-to-consumer drug advertisements. If the organization prevails it will be good news—even though as a journalist I may have second thoughts. Direct-to-consumer drug ads are one of the biggest sources of advertising revenue for many publications.

These advertisements are bad enough, but the Food and Drug Administration and others concerned about public health face something even worse, the promotion of prescription drugs on social media, often by celebrities. For one thing, postings on social media move quickly and change quickly—far outrunning the agency's ability to keep up, even though it has established a "Bad Ad" program that takes tips about ads on social media.[50] Unfortunately, it's directed at medical professionals. It's members of the general public who need to know about the problems with these drugs.

Also unfortunate is the fact that many people would rather take the word of, say, Kim Kardashian West than consult a competent medical practitioner for information about drugs. That's why the maker of a morning sickness drug was so pleased when she took to the internet to plug the product.[51]

The business of medicine feeds on our search for a quick fix, whether medical or surgical. People with back pain, heart disease, type 2 diabetes, or other problems don't want to learn their problem relates to their excess weight, lack of exercise, or aversion to vegetables. And they certainly don't want to hear that, for the moment anyway, medical science has nothing to offer against it.

"The Final Operation"

At one time, autopsies were considered central to the practice of medicine. Until roughly the 1950s, about half of us were autopsied—physicians were often required to perform the procedure as part of their training. That rule is no longer in effect, and requirements that hospitals autopsy 20 or 25 percent of their cases went by the boards decades ago. Today, the autopsy rate is well under 10 percent, and that figure includes people who die by violence or in accidents, when autopsies may be required by law—and are ordinarily paid for by state or local authorities—and at the few teaching hospitals where the practice is emphasized.

The autopsy, "the final operation," as my physician colleague Larry Altman called it in a column years ago, commonly produced interesting findings.[52] Autopsies of auto crash victims led to changes in safety devices. Autopsies of victims of what had been called "cafe coronaries" disclosed that many had choked to death, knowledge that impelled the development and use of the Heimlich maneuver. Autopsies can disclose that a person had a genetic condition, something relatives might need to know about.

"Autopsy will reveal a previously undiscovered major diagnosis in up to 40 percent of cases, a number that has been constant for the 60 years such clinicopathologic correlations have been recorded," Darin L. Wolfe, a pathologist at hospitals in Indiana, wrote in *American Scientist*.[53] That number may be artificially high because cases in which the cause of death is uncertain are more likely to be referred for autopsy. Still, he said, autopsies demonstrate that "the diagnostic error rate remains a very significant medical concern." The misdiagnosis rate is particular high among the elderly, for whom the autopsy rate is generally low.[54]

According to Dr. Wolfe, more than a third of autopsies show major errors. This figure does not necessarily reflect incompetence on the part of physicians. Some conditions may present atypically, lie among multiple other problems a patient has, or simply fail to produce symptoms. Even if they do not discover error, autopsies can provide valuable information about the success of new treatments or diagnostic tests, and they can be tremendously valuable in the education of physicians.

Of course, there are many reasons why the autopsy rate might drop, apart from a reluctance of anyone to pay for them and a fear of what they might turn up. Today, doctors rely more on imaging tests, blood tests, and other measurements, "treating the numbers, not the patient," as a doctor friend of mine says. Overreliance on test results can lead to patient harm and even death. Regular autopsies might illuminate test-driven diagnostic misfires.

In recent years, pathologists have invented noninvasive techniques to study the bodies of the dead, like needle biopsies of major organs, MRIs, and other scans. (Autopsy MRIs produce excellent images, Dr. Wolfe noted, "due to the stillness of the corpse.") Though they are of little use in poisoning cases, for example, MRIs combined with computerized tomography can produce highly detailed three-dimensional images of a corpse's innards.

But all of this work costs money. Medicare stopped paying for autopsies in 1986, and hospitals are unwilling to absorb the cost of the procedure and associated paperwork.[55] And few institutions want to pay for something that has a decent chance of uncovering errors or even malpractice that might otherwise go undetected.

QUESTIONS TO ASK

Many people, including me, find it difficult to challenge their doctors about treatments they propose. And, if we ask doctors for information about treatments they suggest, we may find their answers incomprehensible. But here are some questions to think about when you are considering any kind of medical regimen:

- How does it work? That is, what is the mechanism?
- What are the benefits of the treatment?
- What would happen if I refused?
- What are its side effects?

- If the treatment has potentially severe side effects, is there another effective treatment?

Once the FDA has approved a drug for marketing for a particular condition, physicians may legally prescribe it for anything. Often, insurance companies do not cover this "off-label" use of a drug, but many conditions can be successfully treated by off-label drug use. So if your doctor prescribes an off-label use, ask:

- What is the drug approved for?
- What is the evidence it will help me? (Ask for specifics.)
- Will insurance cover it?
- Are there drugs approved for my condition? If so, why not use them?
- What is the dosage? If the drug is not approved for your condition, it may be best to start with a low dose.

If your doctor recommends a test, consider the recommendations of the U.S. Preventative Services Task Force, which evaluates the merits of all kinds of treatments, tests, and other medical practices, and whose reports are widely regarded as sober and reliable. (They are often attacked by interest groups when they undermine the groups' goals, but that is another matter.)

Don't assume that if your doctor recommends a test you must have it. Sometimes medical tests are merely time-consuming and unpleasant. Sometimes they are potentially dangerous. For example, Americans are exposed to much more radiation from medical tests today than they were a few decades ago.

Avoid tests whose rationale is not clear. In a recent column, Dr. Keith Roach, whose feature "To Your Good Health" runs regularly in newspapers around the country, offered a worst-case scenario of the results of an unnecessary test. "I know of a man who had a non-recommended screening test, which led to a biopsy, which led to an infection of the spinal cord, which ultimately left him a paraplegic."[56]

On the other hand, in her column in the *New York Times,* my colleague Abigail Zuger, also a physician, recounted what happened when one of her patients was (wrongly, she felt) subjected to a CT scan. The scan turned up a lung cancer that was successfully removed.[57]

So when your doctor recommends a test, here are some questions to ask:

- Will the result of this test change my care?
- If the answer is no, why do I need the test?
- Are there alternative methods of obtaining the information or less invasive or damaging tests?
- Are your previous test results good enough? Keep track of your test results and keep copies of the findings.

You may also want to investigate whether or not your doctor has a financial interest in your being tested. Does your doctor own the machine that will be used to do the test, or have an interest in a commercial testing center? According to *Consumer Reports,* doctors who have this kind of financial interest recommend these tests "substantially more" than those who do not.[58]

In the case of MRIs, CT, and nuclear medicine imaging like PET (positron emission tomography), make sure that the facility is accredited by the American College of Radiology or a similar organization.

Request the lowest radiation dose possible. Some tests involving radiation may be personalized to your physique. A thin person may require less radiation than a heavy person. Presumably your radiologist will make this adjustment as a matter of course, but it is good to check. If your doctor says a higher dose will produce a better image, ask how important for diagnosis or treatment the improvement in image quality is.

For both testing and prescription drugs, ignore all advertising. That should go without saying. In particular, use caution online. Drugs, procedures, and regimens can be pushed relentlessly there—both by people with a vested interest or by "patients" who honestly (or otherwise) believe they have been helped by the remedy in question—long before its effectiveness has been demonstrated. Be aware that many supposedly "grass roots" patient organizations are abundantly supported by pharmaceutical companies eager to push their products.

Finally, remember that often it is the patient, not the doctor, who is pushing a particular test or treatment. If you find yourself in this position, step back and review your thinking. If not, you may find yourself on the receiving end of another bit of medical wisdom: *seek and ye shall find.* That is, if doctors looks hard enough, eventually they will find something worthy of medical intervention. Don't invite that unnecessarily.

What's for Supper?

According to some historians, of the Pilgrims who travelled to North America on the *Mayflower,* fewer than a quarter left living descendants. Many of them starved or were so weakened by hunger that ordinary diseases carried them off. Today, the food situation in the United States is quite different. Food is abundant here and relatively cheap. A century ago, plenty of families spent half their household income on food; today, average families spend only about 10 percent of their income on food.[59] If anything, we eat too much. In fact, the World Health Organization reports that in the world as a whole people who are obese are beginning to outnumber people who are food-deprived.

People looking to identify the cause of this vast weight gain have plenty of targets. More of us are in sedentary jobs with no time or energy for exercise. We eat more. If you doubt that, compare classic cookbooks of even a few decades ago with their contemporary equivalents. You will see serving sizes are larger today than they used to be.

Today we see obesity's effects on our lives from the infant car seat (manufacturers make them larger now, to accommodate heftier babies) to the grave (makers of wheelchairs, hospital beds, and even coffins now offer products designed for people who weigh 500 pounds or more). Goliath Casket, "committed to serving the funeral needs of the big and tall," offers a line of what it calls "plus size" coffins.[60]

So food has re-emerged as an issue threatening the nation's health. "[A]bout half of all American adults—117 million individuals— have one or more preventable, chronic diseases, and about two-thirds of U.S. adults—nearly 155 million individuals—are over-weight or obese." That was the conclusion of the 2015 Dietary Guidelines Advisory Committee, established jointly by the Secretaries of Agriculture and Health and Human Services.[61]

The committee attributed this sad state of affairs to poor eating habits on the part of many Americans, overeating generally, and the widespread avoidance of exercise. It noted that Americans obtain their food from a variety of venues but added, "though diet quality varies somewhat by the setting where food is obtained, overall, no matter where the food is obtained, the diet quality of the U.S. population does not meet recommendations for vegetables, fruit, dairy or whole grains, and exceeds recommendations, leading to overcon-sumption, for the nutrients sodium and saturated fat and the food components refined grains, solid fats and added sugars."

In other words, we are avoiding fresh, natural foods in favor of highly refined food "products." As a result, the committee concluded, "obesity and many other health conditions with a nutritional origin are highly prevalent."

What should we be eating? As the food writer Michael Pollan says, we should eat things our grandmothers (or great-grandmothers) would have recognized as food, and most of it should be fruits or vegetables. But a lot has changed since my grandmother was cooking in her New Jersey kitchen. Instead of whole foods like roast chicken, we have processed foods like pre-frozen chicken nuggets. The gardens of the Garden State have given way to sub-urbs. Despite the locavore movement and growth in farmers' markets, we don't eat much locally grown food anymore. Instead, we eat highly processed food manufactured from crops from distant farms, much of it eaten in fast food joints, not at the family dining table.

These trends are good news for agribusiness, because whole foods cooked and eaten at home are not its big profit center. The big profits lie in foods manipulated out of their natural state, often by the addition of fat, salt, and sugar.

This kind of processing is what economists call "adding value." Usually when they use the term they are referring to something like a piece of wood transformed into a piece of furniture, but in this context potato chips are potatoes with value added. Food manufacturers thrive on value-added products. They maintain a healthy bottom line by encouraging unhealthy eating.

For years, one of their major weapons was the so-called "food pyramid," dietary guidance issued by the U.S. Department of Agriculture in the form of a chart with a broad base of grains, hefty helpings of meat; cheese, eggs and other dairy products; and even sugars and sweets. Because of the influence of agribusiness on the Department of Agriculture, we were never advised to "avoid" even things that were obviously not good for us. Instead, we were urged to consume these foods "wisely."

The USDA unveiled a food guide pyramid in 1992.[62] It was not particularly helpful. It called for six to eleven servings a day of bread, cereal, rice, and pasta, which in this country means inevitably high-carbohydrate food; two to three servings of milk, yogurt, and cheese—without reference to whether it should be full- or low- or nonfat; two to three servings of fruits and between three and five of vegetables; two or three servings of meat, poultry, fish, dry beans, eggs, and nuts—all lumped together in one category; and fats, oils, and sweets, similarly lumped together but with the advice: "use sparingly."

So other groups began creating their own food pyramids. Harvard Medical School and the Mayo Clinic produced them. People devised diet pyramids based on peanut butter, cabbage, or grapefruit. They have even devised food pyramids (supposedly) tailored to your blood type.

The Harvard School of Public Health established its own pyramid that put red meats and butter in their own category and advised they be "used sparingly." It said the same for another category, sweets and the high-carb foods like potatoes, white rice, and white bread and pasta that formed the basis of the government's pyramid. The School of Public Health recommended one or two servings of dairy products (or a calcium supplement), between one and three servings of nuts and legumes, zero to two servings of fish, poultry, and eggs (that "zero" was an innovation—not the kind of thing Big Ag ever wants to see in a food guide); plant oils including soy, olive, canola, corn, peanut, and sunflower oil; and vegetables "in abundance."

Because the federal government supports school lunch programs, the Department of Agriculture also has a lot to say about what the nation's schoolchildren eat for lunch and, in many jurisdictions, breakfast as well. Because, overall, we are producing too much milk and other dairy products, the school lunch program, like food aid programs for the poor, lean heavily on products like American ("process") cheese.

Michelle Obama has done a lot to turn this situation around, but critics of the program still assert it relies too much on full-fat dairy products and processed food such as "fruit cocktail," made with high-fructose corn syrup rather than just fruit. Meanwhile, critics on the right deride Mrs. Obama's efforts as officious attempts to create a "nanny state"—even though, possibly because of her influence, rates of childhood obesity seem to be leveling off.

Though many nutritionists—and consumers—applauded decades ago when food companies were required to provide "nutrition facts" about their products on their labels, there is a growing sentiment these days that the labels provide too many data points but too little overall guidance about whether the product in question would be a good choice or, as the food writer Mark Bittman put it in a column in the *New York Times,* "whether something is really beneficial."[63] Bittman advocated labels that rated foods according to their nutrition; their "foodness" ("how close the product is to being real, unadulterated food") and its "welfare," measured in terms of the impact of its production on those who make it and the environment in which it is produced.

By this measure, Bittman said, a jar of organic tomato sauce would receive fourteen out of fifteen possible points, while a box of the imaginary cereal Chocolate Frosted Super Krispy Krunchies ("barely recognizable as food") would come in at four (winning some points for its fiber content).

Advocates of organic agriculture say it is obvious that organic food is better for us than plant or animal food grown with the aid of antibiotics, artificial fertilizers, chemical pesticides, or other interventions. Are they right?

There is abundant evidence that organic farming is far, far better for the environment and far, far more humane to farm animals than so-called "conventional" agriculture. Organic methods can actually leave the land better off than it was before, which is one reason why Amish farms are among the most productive in the United States. And animals raised to graze or forage naturally have a much more pleasant life than those in factory farms. (As an advocate of raising beef on grass once told me, grass-fed cattle "have a really nice life and then one really bad day.")

So it is good to encourage organic farming by buying organic food. Nutritionally speaking, however, there is little reason to believe organic food is better for us. When researchers at Stanford reported in 2012 the results of their studies of organically raised produce, "They concluded that fruits and vegetables labeled organic were, on average, no more nutritious than their conventional counterparts, which tend to be less expensive. Nor were they any less likely to be contaminated by dangerous bacteria."[64] The researchers noted that conventionally grown fruits and vegetables did have more pesticide residue, "but the levels were almost always under the allowed safety limits." Anyway, you can solve that problem by washing your produce—a good idea even (or, possibly, especially) if it was raised by manure-rich organic methods.

For some people, "almost always" is not enough of a guarantee, especially since they don't trust the government's safety limits in the first place. Also, they may be unaware that plants naturally produce a suite of pesticides to protect themselves against natural predators. Many of them act in the human body just the way synthetic pesticides do. And they don't wash off!

Anyway, the researchers, who performed a statistical analysis on 237 studies of organic food—a market estimated by industry organizations and government agencies to be worth as much as $40 billion or more annually—found that other variables, like ripeness, had more to say about how nutritious a given vegetable or piece of fruit might be.[65]

Also, as they noted, organic food is more expensive than conventionally grown food. That is another reason we should all be eating it, to encourage more production and, one hopes, lower prices. Meanwhile, though, parents struggling to afford nutritious food for their children will have to think twice about organic produce. I am not particularly pressed for funds, but I do not want to pay nine dollars for a small bag of arugula, the going price at the organic farm near me. So it's good that, at least as far as your own personal health goes, it's okay to eat conventionally grown produce.

If you want to eat organically grown food, look for foods with the USDA Organic label. Food with this label was produced without the benefit of hormones, antibiotics, genetic engineering, or most synthetic pesticides or fertilizers. Agribusiness has repeatedly challenged the strict standards of this label. According to Marion Nestle, a nutritionist at New York University and perhaps the nation's leading expert on the influence of agribusiness on our food supply, the industry's agitation over the label is a sign of its reliability.

Other labels are less reliable. Terms like "natural" do not necessarily have standard definitions. And farms that tout their "free range" or "cage free" chickens probably keep their flocks indoors for the first few weeks of their lives, a standard practice to protect them from disease. By the time the farmers open doors (often small) to their chicken houses, the birds may be so used to life indoors that they don't venture outside. Anyway, the life of an industrial chicken is short—when all is said and done, a "cage-free" bird may spend little of its life outdoors.

Since the Stanford researchers published their report, others have challenged its assertions, saying that organically grown food is indeed somewhat more nutritious, though they were unable to point to big benefits. And if people on tight budgets stick to organically grown produce, they may end up eating fewer fruits and vegetables altogether—not a good outcome.

Organically raised, grass-fed beef is one exception—a big one—to the idea that conventional and organic foods are nutritionally equivalent. These cattle are typically raised in pastures where they eat what they evolved to eat—grass. By contrast, conventionally raised beef cattle spend much of their lives crammed into feedlots where they are fattened on corn. Corn-fed beef sounds wholesome, but it isn't. Cattle are ruminants and when they are fed corn they fatten more quickly, but they are more prone to illness. So corn-fed animals are routinely treated with antibiotics to stave off infection, a practice that contributes to the growth of antibiotic resistance.

Corn-fed beef is fattier than grass-fed beef. Estimates vary, but a four-ounce serving typically has between fourteen and sixteen grams of fat, compared to between seven and ten grams in a similar cut of grass-fed beef. Because fat makes things taste good, many people prefer corn-fed to grass-fed beef; they may describe grass-fed beef as having an unpleasant, gamy taste. (That is one reason why many Americans have grown to prefer native lamb to lamb imported from New Zealand, where it is typically grass fed.) Grass-fed beef also has more omega-3 fatty acids, similar to those in fish and, it is believed, beneficial for the heart.

Another nutritional issue relating to cattle is milk, especially bovine growth hormone given to milk cows to cause them to produce more milk sooner. There has been a lot of agitation about the use of this substance by people who fear they will end up drinking milk contaminated by hormones. They needn't worry. The hormone does not get into the milk. It *cannot* get in the milk. That fact does not prevent some opponents of use of the hormone from declaring that it poses a threat to people.

But, like cattle crammed into feedlots eating corn rather than grass, cows treated with growth-accelerating hormones are more prone to disease. As a result they are also routinely treated with antibiotics. *That* is the reason to drink organic milk—it is kinder to the cows that produce it.

People who oppose the use of bovine growth hormone (for good reasons or otherwise) often insist that milk from treated cows be labeled as such. The dairy industry is against labeling, on the grounds that people will think milk not labeled as coming from hormone-free cows is somehow contaminated with hormones, which it is not.

For many people, it also seems obvious that crops grown with the benefit of genetic engineering, so-called genetically modified organisms or GMOs, must be bad for you. There is no evidence for that assertion either. In the United States, unless you avoid processed food products completely, something few Americans can claim to do, you will be eating GMO food all the time. Most processed food contains high fructose corn syrup or substances derived from soybeans, and most corn and soybeans grown in the United States are genetically engineered, often to be able to survive a drenching in herbicides applied to suppress weeds.

Genetically modified crops may increase yields, they make certain pesticides unnecessary and others easier to apply, and they reduce the need for fertilizer. Some genetically modified crops, like "golden" rice engineered to contain more Vitamin A, could potentially help preserve eyesight in poorer countries.

Once again, the issue is not harm to human health. (Some people worry about the possible importation of allergens into genetically engineered plants, but we have yet to see that.)

Some researchers have asserted that migrating monarch butterflies are harmed when they pass over fields of genetically modified crops treated with heavy doses of pesticides. But that is only one of many threats to the monarchs. Others are climate change and, especially, habitat loss.

Periodically there are efforts to require food producers to label foods containing ingredients made from genetically modified crops. Food makers always fight them on the grounds that labeling serves no health-related purpose and will only frighten consumers away from their products. Vermont has a labeling requirement, set to take effect in 2016, but meanwhile a bill exempting food companies from the requirement has passed the U.S. House. It will be interesting to see what happens if such rules go into effect. Will the nation abandon its manufactured cakes and cookies and breakfast foods and nachos? I doubt it.

In May 2016, an expert panel convened by the National Academies reported that evidence so far suggests that genetically engineered crops are safe to eat and do not harm the environment. The panel was uncertain about whether use of the technology increases crop yields and added that so far few traits—notably resistance to insects, pests, and herbicides—had been engineered into crops.[66] In the end, the panel offered talking points for people on all sides of the issue.

Apart from being the gateway GMO additive, high-fructose corn syrup is also implicated in the vast increase in obesity in the United States since the 1970s. In 2015, federal health officials reported that about 38 percent of American adults were obese in 2013 and 2014, up from about 32 percent ten years earlier and from about 15 percent in 1970.[67] Though the obesity rate among children had not risen much in that time, that finding was hardly good news: 17 percent of them were obese. (According to government standards, a person who is 5'9" tall is overweight at 169 pounds, and obese at 203 pounds.)

Many associate the rise in obesity in the United States with the Nixon administration's decision to alter farm support policies in ways that encouraged production of crops like corn. The result was a glut of corn, which savvy processors turned into a glut of corn syrup, which is much easier to store and ship. Pretty soon high-fructose corn syrup began turning up in products ranging from crackers to imitation maple syrup. It is practically ubiquitous in processed food, a major reason why that kind of food is so tasty and so dense in calories.

People who manage to eliminate it from their diets usually lose weight. So consumers began pressing manufacturers to remove high-fructose corn syrup from their products—in part by posting mocking videos on YouTube and by starting Facebook pages. By 2010 many were beginning to comply.[68]

The bad rap against high-fructose corn syrup eventually led the Corn Refiners Association to ask the Food and Drug Administration to change its name to "corn sugar." Simultaneously, television ads began appearing in which concerned moms who were avoiding products with high-fructose corn syrup were informed that it was just another form of sugar.

And it is. High-fructose corn syrup is metabolized like sucrose—ordinary sugar made from cane or sugar beets. Or, as the *New York Times* put it in an editorial endorsing the name change, "an -ose is an -ose."[69] Natural sugars, like those that occur in dairy products, fruit, and some vegetables, are better for you not because of the sugar itself, but because of the extra nutrients you get from the foods. Nutritionally speaking, brown sugar, raw sugar, white sugar, and high-fructose corn syrup are comparably devoid of nutritional benefits—except calories.

In 2015, Coca-Cola recruited (paid) fitness and nutrition experts to make the case, in on-line posts in February (American Heart Month), that its mini-cans of soda are a nutritious snack—despite the fact that each 7.5 ounce can contains ninety calories, almost all of it from high-fructose corn syrup—the definition of empty calories.

Coke provided money and other support to the Global Energy Balance Network, an organization advancing the argument that people are not paying enough attention to exercise, which, it argues, is the main issue in weight control. After Coke's support became known, the network denied that the $1.5 million it had received from the corporation had any effect on its work.[70] Yet the Associated Press reported that Coke "helped pick the group's leaders, edited its mission statement and suggested articles and videos for its website."

Perhaps becoming aware of an unfolding public relations disaster, Coke conceded that "there was not a sufficient level of transparency with regard to the company's involvement with the Global Energy Balance Network." By then, however, health experts and others were condemning Coke as attempting to persuade people that more exercise could counteract the effects of its own efforts to flood the nation with sugary soft drinks.

Plenty of health experts had long pointed to these drinks as a major driver of obesity. Some enlisted public officials to fight this plague by, among other things, establishing new taxes for the drinks or limiting the sizes of typical servings. The latter approach was defeated in New York City, despite the support of its mayor at the time, Michael Bloomberg. But efforts to tax soft drinks in order to slow their sales seem to have succeeded in Mexico where, in 2013, lawmakers passed a tax on sugary soft drinks that increased costs for bottlers by about 10 percent.[71] After a year, soda sales were down by 13 percent; declines were largest in neighborhoods of low-income Mexicans.

As an aside, here's another example of the kind of self-evident truth that turns out not necessarily to be true: drink diet soda if you want to lose weight. Diet drinks are typically made with aspartame, saccharine, or other calorie-free sweeteners. But no studies have confirmed that people who eat foods made with these sweeteners lose weight. Some research even suggests that people who consume things like artificially sweetened diet soda actually gain more weight than those who don't. The reason so far is unknown, although there are a lot of theories. Maybe artificial sweeteners stimulate the appetite in some as yet unknown way. Maybe they interfere somehow with the action of insulin in the body. Maybe they affect the suite of bacteria that normally colonize our gut, our so-called microbiome. Maybe people who drink diet soda feel freer to consume calories in other ways.

In May 2016, Michelle Obama unveiled changes to the FDA's "nutrition facts" label found on most grocery items. The new label gives more prominence, in the form of larger type, to the number of servings a package contains and the number of calories per serving. In theory, these changes will make it more likely that consumers will notice that food servings have relatively few calories only because the serving size has been defined at an unreasonably low level. (Who eats only half a cup of ice cream?)

The label must also list "added sugars," i.e., sugar not present in the food before the product was processed or manufactured, *and* how much of this added sugar people ought to consume every day. The label decision was a disappointment, said the Sugar Association, a trade organization whose mission, according to its website, is "to promote the consumption of sugar through sound scientific principles."[72] Still, U.S. sales of high-fructose corn syrup have begun to fall. Big producers are shifting their marketing focus abroad.

Meanwhile, food companies are using advertising to tout the supposed health advantages of everything from roast beef to pomegranate juice. Some of the most effective ads come from the dairy industry, whose products are no longer so easy to fit into an officially "healthy" diet. (And the question of which dairy foods to consume cannot accurately be answered with a phrase like "Got Milk?" or "3-A-Day," both slogans that have been adopted by the dairy industry.)

In general, health claims made for processed foods must be viewed with suspicion. Since 2006, when the European Union began requiring food makers to meet strict scientific standards if they wanted to make health claims about their products, the European Food Safety Authority has rejected 80 percent of proposed claims. Apparently, some manufacturers were citing Wikipedia, the dictionary, or even the Bible in support of their products. "The simple act of asking for evidence is sometimes enough to reveal the shoddiness of a claim."[73]

In the United States—unlike most other developed countries—such health claims are made frequently and may even be directed at children. According to the Rudd Center for Food Policy and Obesity at Yale, the average child in the United States sees more than a dozen food commercials on television every day (4,700 per year) while teenagers encounter sixteen per day (5,900 per year). By contrast, the center said, children see on average one ad per week for foods like fruits and vegetables.[74] Marketing experts say these ads heighten children's "pester power" and turn them into drivers of decisions parents make in the grocery store.

YouTube Kids, an online channel for children, has drawn complaints from advocacy groups like the Campaign for Commercial-Free Childhood and the Center for Digital Democracy, which argue that the videos on the site amount to little more than advertising, and without the benefit of Federal Trade Commission scrutiny at that. The groups also charge that YouTube is violating its own policies for limiting advertising to children.

According to the Federal Trade Commission, 2009 data from forty-eight food companies showed those companies spent $1.79 billion annually in marketing to "youth." "Companies continue to use a wide variety of techniques to reach young people, and marketing campaigns are heavily integrated, combining traditional media, internet, digital marketing, packaging, and often using cross-promotions with popular movies or TV characters across those. These techniques are highly effective."[75]

The World Health Organization supports regulations to reduce the exposure of young people to advertisements for food products. But the policies of the WHO are not immune from criticism. (Think of its supposed warning on the supposed brain cancer risk of cell phones.) Its declaration that a substance is a possible carcinogen means only that some amount of exposure is associated with some level of increased risk.

The agency ran into a storm of criticism in 2015 when it classified processed meat, including hot dogs, as a carcinogen. Just two strips of bacon a day could raise a person's risk of colorectal cancer by 18 percent, the agency warned. What few people noticed was that, even if the agency's assessment was correct, eating bacon daily would raise one's lifetime risk of the disease from about 5 percent to about 6 percent.[76]

Some people tout aquaculture as a possible way to feed the world's growing population, especially now that more people want—and can afford—to make animal protein a regular part of their diets. Aquaculture's record is spotty, though—again, not because of what happens to people who eat farmed fish (nothing untoward, apparently) but rather, because of what fish farming does to the environment. Salmon is a good example.

Most wild stocks of Atlantic salmon have long been fished into commercial extinction. That is, there may be some of the fish around, but not enough to produce a catch large enough or reliable enough to support a fishery. Nowadays, virtually all Atlantic salmon at the seafood counter is farmed in Europe, Chile, or off the coast of British Columbia. These fish are hatched in mainland hatcheries and raised in near-shore pens containing tens or even hundreds of thousands of fish.

These farms have many problems. The fish in the pens generate prodigious quantities of fish poop, which falls to the bottom, creating areas of pollution. Fish crowded into pens are subject to diseases, which lead many growers to treat them routinely with antibiotics, adding to the problem of antibiotic overuse. Also, diseases and parasitic infection incubated in the crowded pens can spread to wild stocks.

Perhaps worse, from the environmental perspective, the salmon in these pens are carnivores and they have to be fed. Usually they are fed pellets made from other fish, like herring. Efforts to feed farmed fish are depleting stocks of these feed fish, which occupy an important place near the base of the marine food web. So far, fish farmers have yet to break even when it comes to salmon because it still takes more than a pound of fish processed into feed to produce a pound of salmon. Also, because the fish food pellets do not closely mimic the diet of wild salmon, the flesh of farm-raised fish is not the "salmon" color of wild fish. So producers introduce either synthetic or naturally occurring astaxanthin—the compound that makes shrimp pink, for example—into the feed to transform what would otherwise be pale, grayish fillets into something with a more appetizing "salmon" color.

Is any of this bad for us? No. There was a modest hoopla a few years ago over the supposed harm to people who ate farm-raised salmon, but the risk, if any, is minuscule, overwhelmed (if it even exists) by the large health advantages of eating salmon in the first place.

But we should strive to avoid farm-raised Atlantic salmon in favor of wild-caught Pacific salmon. Can you trust the labels in your fish market? When the conservation group Oceana surveyed restaurant and fish market salmon sold as wild-caught it found two-thirds of the restaurant salmon and 20 percent of the market salmon was actually farmed.[77] Like organic produce and grass-fed meat, eating wild-caught salmon is more expensive, but it is much better for the environment.

The same is true of shrimp. If you can afford it, and you can find it, eat wild-caught shrimp. Unfortunately, much of what is on sale in American groceries is farmed shrimp, often from Asia, raised in what once were coastal mangrove forests. These mangroves, crucial for the region's ecology and for storm protection, were flattened to make way for the aquaculture establishments. There are even allegations that shrimp from Asia are processed using slave labor.

Many people worry about mercury contamination in fish. Mercury accumulates in some species, like tuna and swordfish, at the top of the food web. In my opinion, people worry too much about it—not because mercury in your tuna is good for you, but because the risk of consuming it in fish is outweighed by the dietary benefits of the fish. For most of us, the major source of mercury is air pollution from coal-fired power plants (the source of the mercury contaminating the fish). But, as with particulate pollution, no one is afraid of coal-fired power plants. There is abundant evidence that eating fish is good for us, and that most Americans could benefit by eating more of it. By some estimates, as few as 20 percent of us eat the recommended two servings of fish per week.

Though environmentalists complain about the regulation of fishing in this country, by international standards most of our fisheries are relatively well run. If you want to eat fish but worry about sustainable fishing, consult the website of the Marine Stewardship Council, a group based at the Monterey Bay Aquarium, in California.[78] The council tracks the health of fish stocks around the world and identifies fish and shellfish "harvested in a sustainable manner."

If your goal is weight loss, there is no shortage of people ready to offer diet advice. But it is extremely difficult to subject diet plans to the gold standard of scientific research, the placebo-controlled, double-blind clinical trial. Once in a while, researchers will acquire enough money to recruit subjects who agree to be locked up for weeks while technicians track every morsel that enters their mouths and every bit of physical effort they exert. But this kind of work is expensive, and lining up willing participants is difficult.

Instead, researchers who study diet monitor what people eat by having them keep food diaries and otherwise track what they consume, a far less reliable method. Also, few studies of any kind of weight loss approach last more than eighteen months. Critics of various diet programs say that is one reason why their high failure rates are little known—the studies end before many participants have time to relapse. By some estimates, more than 90 percent of people who successfully diet gain all the weight back within a year or so—sometimes even more.

Over the years, people with one kind of credential or another, or no credentials, have advanced every possible kind of diet: low fat, low carb, vegetarian, vegan, and paleo, along with diets based on cabbage soup, grapefruit, fermented foods, and so on. One thing almost all of them have in common: someone is making money on them.

One fad is the paleo diet, based on the idea that we should eat the foods our Paleolithic (Stone Age) ancestors ate—nuts, berries, and other plants (uncooked), and meat. As people like Michael Pollan note, unless we adopt the hunter-gatherer lifestyle and find a landscape in which we can hunt and gather with success, we won't be able to duplicate the paleo lifestyle.[79]

Also, raw food is not necessarily better for you. While it is true that you can boil the life (and the nutrients) out of some vegetables, cooking usually increases the nutrients *available* to us in the food we eat. That is why cooking is virtually universal in human culture.

One of the most contentious food issues is fat. For years people were advised—by the U.S. Department of Agriculture and the nation's medical establishment, among others—to avoid it. We responded by eating carbohydrates instead, sugars and starches. Eventually it became clear that we had, in effect, made a bad situation worse.

In part because of the low-carb craze popularized by Robert Atkins and others, it is now clear that fat per se is not by definition bad for one's health. (Though many credit Atkins with inventing the high-fat/low-carb diet, its first advocate seems to have been a man named William Banting, an obese nineteenth-century British undertaker who used the diet in an attempt to lose weight.) Monosaturated and polyunsaturated fats, found in foods like fish, olives, and avocados, seem to be benign, and even the saturated fats of butter, cheese, and red meat are okay in moderation.

But in the 1970s, that view would have been regarded as heresy. Doctors advised patients to remove fat and especially cholesterol from their diets. Food manufacturers, eager to rid their processed products of fat, filled them instead with carbohydrates. Now it is becoming clear that carbohydrates are implicated in a host of serious health issues, including diabetes and heart disease.

Does the low-carb approach work? Advocates swear by it. But according to newspaper accounts, when Dr. Atkins died at 72, apparently of heart disease, he weighed 258 pounds.[80]

Anyway, the levels of fat and cholesterol in one's diet are surrogate endpoints. What we care about is not cholesterol in the diet or even, really, cholesterol levels in the blood. It's the incidence of heart attack and stroke and other vascular problems. And the connection between dietary cholesterol and blood levels of cholesterol is tenuous, probably because cholesterol—an essential compound for essential organs including the brain—is produced by the body, primarily in the liver, and what goes on there may be more significant, clinically, than what we eat.

One thing we have learned: focusing on one class of foods as a dietary villain is a recipe for disaster. At the moment, what looks good is a balanced diet, heavy on vegetables, light on meat, and as free as possible of processed food.

In any event, the government abandoned its dietary pyramid schemes in 2011, replacing the latest version with a symbolic dinner plate marked off in servings—most of them fruits and vegetables.[81] It is hard to say how much influence these new "plates" will have. Few people consult government recommendations when they plan menus or go out to a fast-food restaurant for dinner.

One thing is certain: the government is not putting its money where its dietary mouth is. When it comes to agricultural subsidies, meat and dairy come first, followed by grains. Vegetables and fruit bring up the rear.

In February 2015, the Dietary Guidelines Advisory Committee, an arm of the federal Department of Health, issued food advice that changed the fat-related playing field, at least a bit. Among other things, the committee said that dietary cholesterol was not the bête noir it had been described as, though it said people should eat less red and processed meat.

The reaction was predictable, and it came from all sides. The North American Meat Institute, which bills itself as the oldest and largest organization of meat packers, protested that its products were being demonized. "We've been put in a position over the years to almost be apologizing for our product, we're not going to do that any more," its president and CEO, Barry Carpenter, told the Associated Press.[82]

The institute also supports Meat MythCrushers, an effort "to provide consumers and media with the other side of the story."[83] And it enlisted their allies in Congress (most of them Republican, the Associated Press reported) to combat what one of them, Michael

Conaway, called "flawed recommendations." Conaway, a Texas Republican, heads the House Agriculture Committee.

Many of the people who offer dietary advice—even academics—are so tainted by commercial ties, consulting agreements, and research grants that it is hard to take their guidance to heart.

After the federal dietary guideline committee's report came out, Nina Teicholz, author of a book touting the dietary advantages of butter, meat, and cheese, wrote a commentary in the *New York Times* about the panel's suggestions. Though some will disagree about her particular conclusions, her overall thesis is hard to dismiss. For decades, she wrote, "Americans have been the subject of a vast, uncontrolled diet experiment with disastrous consequences. We have to start looking more skeptically at epidemiological studies and rethinking nutrition policy from the ground up."[84]

It is possible to worry so much about eating right that you develop "orthorexia nervosa"—an unhealthy obsession with eating only healthy foods. The term, apparently coined in 1996 by a physician, Steven Bratman, is not included in standard directories of mental illness—yet. Here are some suggestions for avoiding it:

- Eat a balanced diet, heavy on the vegetables and fruits. Avoid supplements unless your doctor or nurse recommends them.
- View with suspicion any health claim for any food, but particularly processed foods. We are barraged with these claims in advertising. Remember, the people making them are attempting to sell the product.
- Read labels. Reject foods if you need a degree in chemistry to know what you will be eating. Avoid foods whose first listed ingredients—the substances most prominent in the food—are fructose, high fructose corn syrup, or other sugars.
- Eat whole foods, organic if you can find them and can afford them.

- Start to think of meat as a bit player, rather than the star of your meals. If you can find it and you can afford it, try grass-fed meat.
- When you eat out, even in fast-food joints, patronize establishments that avoid meat produced with antibiotics and look for menus that tell you how many calories you will consume if you eat what turns up on your plate.
- When you buy seafood, ask the fishmonger where it comes from. If the seller cannot give you a good answer, try another shop!

In a talk at Brown University in 2008, Michael Pollan offered another bit of advice: don't eat anything that doesn't eventually rot or get stale. In his office, he said, he keeps a Twinkie—as soft as the day he bought it years earlier.

5

POLITICAL SCIENCE

Constituency of Ignorance

Many researchers believe that if policy makers knew as much as researchers know, they would make policy researchers would like. When things don't work out that way, the researchers assume ignorance (or worse) on the part of policy makers is to blame. People who study science communication call this thinking "the deficit model." It does not come close to explaining what goes wrong when science and politics meet.

Of course, sometimes ignorance *is* the issue. There are government officials and policy makers who are ill-informed or uninformed. Some of this ignorance can be attributed to researchers who, overall, make few if any efforts to explain their world to the public or to public officials and who, as a group, disdain the hurly burly of politics. Many think colleagues who engage politically have somehow tainted themselves or abandoned the objectivity of the lab.

So it is not a surprise that fewer than 10 percent of members of the House and Senate in the 114th Congress of the United States (in office in 2015 and 2016) were scientists or engineers.[1] That's not very many, especially when you realize that figure included the optometrist in the Senate and seven software company executives (two in the Senate and five in the House). There were eight engineers, all but one in the House, which also boasted a physicist, a microbiologist, and a chemist. The others in this group were health care providers: eighteen physicians (all but three in the House) and four nurses, three dentists, three psychologists, and a pharmacist—all in the House.

By far, the professions most heavily represented were public service (i.e., politics), law, and business. (The research service says there were also nine journalists, all but two in the House. And one senator, the report notes, was "a comedian." Presumably that is Al Franken, the Minnesota Democrat.)

But ignorance of science is only one of many problems that stand in the way of making sensible policy in government.

In the first place, many questions that people describe as issues of science are not science questions at all, but rather questions hinging on deeper beliefs and values. Some examples are obvious. Your views on the therapeutic use of human embryonic stem cells are likely to rest on your ideas of the moral status of the human embryo. Others are less obvious. People think that the climate question is one that science can answer. But in a larger sense, the answer depends on whether you think people living today have obligations to leave Earth more or less intact for generations after us. That is not a question science can answer.

Politics and policy making are not a search for the technically correct answer to this question or that. They involve debates over advancing our social aspirations—social welfare, national security, support for the arts, economic prosperity—through a process of negotiation, log-rolling, and political action. But politicians understand that Americans have great respect for scientists and engineers. So they say, "the science is on my side."

Sherwood Boehlert, a Republican who for many years represented a district in upstate New York in Congress, offers a good example, congressional consideration of a Clinton administration proposal to tighten restrictions on ground-level ozone. At the time, he chaired the House Committee on Science, Space, and Technology, the go-to committee on science in Congress.

In the upper atmosphere, ozone offers a crucial protection from dangerous ultraviolet radiation. When scientists realized that ozone from aerosol sprays and certain other products was making its way to the upper atmosphere and eroding a hole in this ozone layer, the nations of the world acted to ban or greatly restrict their use, through a 1987 agreement called the Montreal Protocol.[2]

At ground levels, though, ozone pollution produces what we know as smog. And it sends people with asthma or other breathing problems to the hospital in great numbers. Excess hospital admissions and even deaths track neatly with ground levels of ozone.

But requiring businesses to take steps to limit it would be costly. So—where to draw the line? What should the limit be? People thought of that as a science question and, Boehlert related, many of his colleagues in the House asked him what the "scientific" level would be.[3]

Unfortunately, as Boehlert noted, the question was not one science could answer. The issue was not ozone per se but rather a far stickier issue: How much illness and how many deaths from ground-level ozone pollution are you willing to accept? That was not a question anyone in Congress wanted to confront.

When Boehlert headed the science committee, David Goldston was its chief-of-staff. Goldston, who today is director of government affairs at the Natural Resources Defense Council, did not train as a scientist (he studied American history in graduate school), but he was in the thick of many of the nation's most important science and technology debates for years. He developed a four-part taxonomy of scientific and technical issues that confront policy makers.

- Issues of values, not science. Ground level ozone was a values question. It was not a question science could answer.
- Issues, like climate change, on which the science is settled but a die-hard (and possibly noisy) remnant refuses to acknowledge it. The vaccine debate is a similar issue. It is settled science that childhood vaccinations do not cause autism, but still some misguided parents persist in demanding their children be exempted from school vaccination requirements.
- Issues on which there is legitimate scientific debate, like whether a low-fat or a low-carb approach is the best way to avoid heart disease.

- Issues that are so new that research has barely begun. The safety of nanotechnology, particularly the production of nanomaterials that make their way into the environment, or even into our bodies, is an example. So is geoengineering, the potential use of engineering techniques to "tune" Earth's climate.

In politics, people do not necessarily pay attention to such distinctions. And when they confront scientific or other technical information that might conflict with dearly held business goals or ideologies, they have a number of ways to quash it.

First, they may deride it. Many creationists, for example, deride evolution as "just another theory." Climate deniers describe researchers in the field as somehow perpetrating a lucrative hoax.

They may target individual scientists for "investigation." The most famous victim of this tactic was Michael Mann, the climatologist at Penn State whose description of the rising curve of Earth's temperature has become known as "the hockey stick." In a disgraceful episode, Republicans in Congress tormented Mann with subpoenas for emails, lab records, and other documents, and demands for testimony, until a specially convened panel of the National Academy cleared him of any scientific wrong-doing.

Politicians may pack scientific advisory committees with people appointed for their political or religious views rather than their scientific or engineering expertise. This practice is not limited to Republicans, but it was a hallmark of the administration of George W. Bush and may be a Trump tactic as well.

More dangerously, they may claim that the situation is uncertain or cook up controversy where none exists. This tactic has become one of the most effective weapons in the arsenal of those seeking to elude government regulation. The Surgeon General declared in 1964 that smoking cigarettes produced lung cancer and other harms, but cigarette company executives banged the uncertainty drum relentlessly for decades, asserting that there was not enough evidence to say what epidemiologists had been saying for half a century.

Climate deniers invoke uncertainty to say we don't know yet whether acting to reduce emissions of greenhouse gases is worth the widespread economic disruptions they say a carbon tax or other measures would cause. In doing so, they magnify the uncertainty over human-induced climate change—there *is* some, but not any that matters in this context—and they magnify the certainty that shifting to a carbon-free economy would be an economic disaster.

The most pernicious maneuver in the realm of science and politics is the deliberate, flat-out suppression of information by factions that find it politically unpalatable. Gun control is an excellent example. About a third of American households are armed, and guns cause 30,000 deaths in the United States each year. Given these figures and gun control's prominence as a subject in so many political debates, you might think that gun safety and related issues would be matters of robust ongoing research. But they are not. For more than two decades the National Rifle Association has led a successful campaign to bar the use of federal research funds for studies of gun violence.

The effort took off in 1993, after the *New England Journal of Medicine* published a research report, "Gun Ownership as a Risk Factor for Homicide in the Home."[4] The researchers had studied almost 2,000 homicides in three urban counties, almost a quarter of which had occurred in the home of the victim. After controlling for factors like illicit drug use in the home, the presence of a person with an arrest record and other factors, they found that, far from conferring safety, "keeping a gun in the home was strongly and independently associated with an increased risk of homicide." They added, "virtually all of this risk involved homicide by a family member or intimate acquaintance."

The work was led by Arthur L. Kellermann, at the time a professor at the medical school of the University of Tennessee and now the dean of the medical school at the Uniformed Services University of the Health Sciences, sometimes referred to as "the nation's medical school."

In the years since, some have criticized Kellermann and his colleagues, saying among other things that they cherry-picked their data. But we don't know if their results hold up, because their research has not been replicated. In the wake of their paper's publication, the National Rifle Association began a campaign against the agency that financed the work, the National Center for Injury Prevention, an arm of the Centers for Disease Control and Prevention. In 1996, Congress passed an appropriations bill explicitly barring the use of injury prevention funds "to advocate or promote gun control." The measure cut the agency's budget by exactly the amount the agency had spent on gun-related research the year before.

"Precisely what was or was not permitted under the clause was unclear," Kellermann wrote later. "But no federal employee was willing to risk his or her career or the agency's funding to find out." Funding dried up, he wrote, and "even today, 17 years after this legislative action, the CDC's website lacks specific links to information about presenting firearms-related violence."[5]

When researchers at the National Institutes on Alcohol Abuse and Alcoholism embarked on related research, Congress extended the restrictive language to all agencies of the Department of Health and Human Services, including the National Institutes of Health.

In the wake of the Newtown, CT school shooting, President Barack Obama called for research on gun violence and money to support it. But he added that the work must be done "in accordance with applicable law." And because the NRA opposes any research it regards as "politically slanted," we may never know whether the presence of a gun in the home really makes it more or less dangerous.[6]

One federal agency, the National Tracing Center, tracks guns used in crimes, noting who purchased them and when. It handles almost 400,000 trace requests each year.[7] The data would be useful to would-be researchers on gun use. But the center, an arm of the Bureau of Alcohol, Tobacco, and Firearms, is allowed to share the data only with law enforcement agencies.

Even though there is little government funding of gun research, some studies have been conducted. Their results are hard to summarize. For example, the influence and effects of the ban on assault rifles, passed by Congress in 1994, is hard to assess, because the law (now lapsed) had no effect on assault rifles already in circulation.

On the other hand, some studies have produced striking results. For example, what researchers call "a natural experiment" played out in 2007, when Connecticut began requiring permits for handgun purchases, and Missouri, which had such a requirement, abandoned it. The result, researchers found, was 15 percent fewer gun suicides in Connecticut and 16 percent more in Missouri.[8]

But research is thin on the benefits of efforts like broader background checks for would-be purchasers, requirements that gun owners report lost or stolen guns, or safety technologies like fingerprint recognition to prevent children from firing weapons. This deliberate embrace of ignorance is not new. I first encountered it when I began writing about coastal ecology for the *Times,* in particular when I was reporting what turned out to be my first article to appear on Page One. The occasion was a hearing on a proposal by the Federal Emergency Management Agency, or FEMA, which administers the National Flood Insurance Program, to revisit flood insurance rate maps, with an eye to including data on erosion risks. When the program began in 1968, the FEMA scientists said, they did not have good data on long-term trends in erosion. But twenty-five years later, improvements in computer software enabled them to say with much greater precision where the coast was eroding, and how fast. The researchers proposed identifying ten-year, thirty-year, and sixty-year erosion hazard areas and altering the insurance rates to reflect this risk.

Though the proposal was eminently sensible, it ran into a buzzsaw at the hearing because, had it ever gone into effect, it would have had a devastating effect on property values at the coast. Few people will be sanguine about buying a house the government tells them has a good chance of falling into the ocean sometime in the next ten years. Plus, as one FEMA official put it at the hearing, an actuarially sound annual flood insurance premium for a house in a ten-year erosion zone would be 10 percent of the value of the house. That would mean a $25,000 premium for $250,000 worth of house.

Perhaps the scientists were naive to think that the overpowering logic of their proposal would overcome the forces of commerce. In any event, they were stunned by the violent reaction against their proposal, which was soundly defeated. "There is a constituency of ignorance on the coast," said Stephen P. Leatherman, then director of the coastal research laboratory at the University of Maryland.

Eventually, even the Congress came to realize that the flood insurance program was an unacceptable drain on the federal treasury, and in 2012 the Biggert-Waters Flood Insurance Reform Act became law. It tightened construction standards and raised insurance rates, among other changes. When Superstorm Sandy struck that fall, the need for the changes became even clearer. But what also became clear was that the new law was highly unpopular along the coast. Legislators began to erode it. Meanwhile coastal flooding gets worse and worse.

Perhaps the most flamboyant example of politics mixing badly with science involved the so-called "Star Wars" plan to base antimissile weapons in space. It was the brainchild of Edward Teller, the physicist who was the model for the deranged Dr. Strangelove of the Kubrick film and who claimed to be the father of the hydrogen bomb.

He nurtured the idea of protecting Americans against surprise missile attack. Until he died in 2003, he insisted that a network of radars, space-based lasers, and other weapons could detect incoming enemy missiles and destroy them in the atmosphere, before they could harm anyone on the ground. He sold this dream to a number of politicians, especially Presidents Ronald Reagan, George H. W. Bush, and George W. Bush.

It is a beautiful idea. But it has a major flaw: it won't work. Such a system, even if it could be built, would always be easy to outwit by simple measures like sending up lots of decoys, or surrounding an incoming missile with radar-blinding chaff, or even over-whelming the defense with sheer numbers of attacking missiles. From the get-go, there was a strong consensus among weapons experts and physicists that it would be a mistake to stake the nation's survival on this kind of approach.

Early tests of prototypes were disappointing. One of Teller's original ideas—a system of space-based lasers—proved impossible to construct. Tests of rockets meant to shoot down incoming missiles showed they could not differentiate between real weapons and harmless decoys. Even when the prototypes were programmed with explicit information about where the attacking missiles were coming from, when they would be launched and what they looked like—their heat signatures and other hallmarks—the defense system could not "see" them. But for advocates of the system, failure was not an option. So every time the system failed, the tests were made easier. When that approach proved inadequate, designers changed the definition of success.

Most scientists and engineers kept silent—but not all. The physicist Robert Park, in his book *Voodoo Science,* described Teller as a man with an "almost unblemished record of failure" and "brilliance untempered by judgment." He dismissed one of Teller's proposals—the so-called "brilliant pebbles" plan to launch clouds of metal chaff to confuse incoming missiles' radar—as "lost marbles."[9]

The space-based ideas were dropped at the close of the Cold War. There are those who say the government never intended to go ahead with the Strategic Defense Initiative, as the proposal was formally known. They suggested it was a ploy to frighten the Soviet Union by suggesting it would have to spend billions to keep up with this new contest in the arms race, and that the fall of communism is attributable in no small part to Star Wars. That will be for historians and Marxist economists to sort out.

Ironically, the demise of the Cold War led to renewed enthusiasm for another project widely regarded as a space-based boondoggle, the International Space Station.

The rationale for the station was never particularly robust. Most of the scientific experiments conducted on the ISS can be accomplished well or even very well in other ways. Some say its principal merit was the way it offered employment to rocket scientists from the newly defunct Soviet Union who might otherwise have sold their expertise on the open market.

Even the space program, which thrilled people like me who watched it unfold as children, was never the science program it was sold as. It was an engineering tour-de-force, especially given its origin in the pre-computer age, but as the writer Daniel Greenberg puts it, "Many scientists considered the moon program a political stunt."[10]

Then there was the Space Shuttle program, which never fulfilled its supposed mission: low-cost transport of satellites to Earth orbit. The necessity of making the shuttle safe for its human crew raised its costs dramatically, and it never met the enthusiastic launching schedules set by NASA each year.

Since the end of the Cold War, there has been something of a decline in the government's in-house scientific infrastructure. For example, Congress's Office of Technology Assessment was abolished in a round of budget cuts pushed through by Newt Gingrich when he was Speaker of the House. That agency had its critics, who complained it took too long to produce its thorough reports, but its death left Congress with relatively few other resources. The Congressional Research Service and the National Academies, the congressionally chartered organization of scientific, medical, and engineering eminence, both move so slowly that their reports are sometimes moot before they have been issued. Also, expert panels convened by the Academies strive so mightily for consensus that they often produce somewhat mealy-mouthed assessments whose more forceful conclusions relate to the need for more money for research—valid, but hardly the point.

The second Bush administration marginalized the White House Office of Science and Technology Policy, cutting its staff, demoting its chief, and moving its offices from the Old Executive Office Building, next to the White House, to the Siberia of rented quarters outside the White House ambit. Eventually, researchers began referring to the whole operation as "a false front." President Obama brought the office back into the White House. Now Trump appointees talk of abolishing it.

By then, some researchers had begun organizing to bring the voices of more scientists and engineers into public debate. One such group was Scientists and Engineers for America. Its goal was the promotion of sound science in government policies and the support of political candidates with scientific or technical expertise, regardless of their part affiliation. It was chronically short of funds and went into abeyance in 2011.

Meanwhile, two physicists, Representative Bill Foster, an Illinois Democrat, and Representative Vernon Ehlers, a Michigan Republican, were organizing a group that eventually became Ben Franklin's List. The two men were leaving the Congress—Ehlers because he was retiring and Foster because he had been defeated by a Tea Party Republican.

Their idea was modeled on EMILY's List, which supports political candidates favoring abortion rights. (EMILY stands for Early Money Is Like Yeast—it can help get a nascent political campaign off the ground.) They chose Ben Franklin because he was a patriot and a scientist.

Their idea, they said when I interviewed them about the organization, was to encourage scientists, engineers, and others with technical expertise to run for political office. They said their decisions about whom to support would be based solely on technical credentials, not political or other views. The group would be nonpartisan banking on the widespread belief that, as a group, scientists tend to be Democrats and engineers, as a group, tend to be Republicans.[11] But their effort also faltered, especially when Foster decided to run for Congress again in 2012. He was elected—but today he is the only PhD scientist in Congress.

The Political Environment

In 1978, James Gustav Speth, chairman of the White House Council on Environmental Quality, gave President Jimmy Carter some bad news: a group of eminent scientists had produced a report warning that if something were not done to reduce reliance on fossil fuels, particularly in the United States, Earth's climate would be significantly altered, with potentially disastrous effects.

Carter had already run into a storm of criticism, even ridicule, for his advocacy of sweater-wearing energy conservation. Petroleum and coal companies did not want to see reduced sales of their products; they wanted to sell more. Automakers focused not on fuel-efficient compacts but rather on more profitable gas-guzzlers, behemoths that drove their bottom line.

Still, Carter forwarded the researchers' report to the National Academy of Sciences, the government-chartered advisory board of eminent scientists, for evaluation. In six months, lightning speed for the deliberative academy, it replied. The danger was real, its experts concluded. The sooner something was done about it, the better.

Almost forty years later, that is still the scientific consensus. But far too little has been done about it.

Gus Speth has had a distinguished career as dean of the School of Forestry and Environmental Studies at Yale, founder or co-founder of groups like the Natural Resources Defense Council and the World Resources Institute. Today he is a professor at the University of Vermont Law School, renowned for its program in environmental law. But when we met a few years ago, he was looking back on decades lost to inaction. "You could get depressed," he told me. "You could get angry. Your predecessors have left you with this problem that is now deeper and more intractable."[12]

This problem is not insoluble. Technology we could take off the shelf today would start turning things around. Relatively easy conservation steps would save immense amounts of energy. Retrofitting a house would cost money, for example, but it would be recouped in lower utility bills in a few years.

But no one wants to tell Americans they cannot drive big cars or heat their houses to 82 degrees in winter and cool them to 65 degrees in summer. And when Reagan replaced Carter as president, he ordered the Carter-installed solar panels removed from the White House roof. On the stump, Trump called climate change a "hoax."

Given the strength of the evidence that human activity is altering Earth's climate in ways that could lead us to disaster, people have struggled to understand why so many people stand in the way of doing anything about it.

Researchers offer explanations. Some are obvious. For example, there has been a "well-organized and well-funded disinformation campaign that has been waged against climate science for decades," as Richard Somerville and Susan Joy Hassol put it in an article in *Physics Today*.[13] Somerville is a climate expert, and Hassol is an expert on communicating information about climate. Unfortunately, they added in their article, for some people the loudest deniers are "trusted messengers" who deliver well-crafted, simple, clear messages—and deliver them repeatedly. Scientists rarely speak out, and when they do, often their jargon-filled remarks are opaque to ordinary people.

For example, when researchers want to say an effect is intensifying, they say it is enhanced. For them, a positive trend is moving upward; the rest of us think they're telling us it is good. And they don't typically spend a lot of time putting things into context or explaining their math—or even expressing things in pounds and yards rather than kilos or meters. (Yes, the United States should adopt the metric system, but we haven't.)

Worse, as Somerville and Hassol note, scientists have developed "a lexicon of likelihood terms (likely, very likely and so forth) to roughly quantify the probability of particular outcomes. The overuse of such terms gives the impression that they know much less than they actually do." As is typical with plenty of other science or technical questions that are subjects of public debate, the research community speaks in ways that can sow confusion rather than understanding.

The media, in reporting both sides, suggest there is a robust scientific debate, rather than settled science challenged by a few outliers. Meanwhile, deniers frame the climate situation as a contest between the environment and the health of the economy.

In fact, it is not action against climate change that threatens the economy, it is climate change itself. It threatens food production, water supplies, the safety of billions of people—everything on which the global economy depends. It even threatens Donald Trump's Mar-a-Lago resort in Palm Beach, Florida. Still, he is staffing his administration with climate deniers and fossil fuel advocates.

Everything we do—even doing nothing—comes at a cost and exposes us to risk. So when we face a choice, whether as an individual or as a society, we want to know what course will produce the best outcome. Disagreements about what to do in response to risk are common, especially when potential problems are poorly understood but potentially dangerous, or if the risk is hypothetical, but the cost of dealing with it is not. This process is full of uncertainty and the arguments about it are full of spin.

First we must decide what the "best" outcome would be. The longest life expectancy or the longest time free of pain? The clearest air or the fastest transportation? The answers to these questions are not always intuitively obvious. It is not even intuitively obvious that they have a single right answer.

My favorite example of a situation in which the risk is hypothetical (and poorly understood) but the solutions are costly is the so-called Asilomar Conference, organized in 1975 by scientists concerned about the potential hazards of research involving gene splicing or, as they call it, recombinant DNA—the mixing of genetic material from different species. Scores of scientists and also some lawyers and physicians turned out at the Asilomar Conference Center in Pacific Grove, California, to discuss the issue. They agreed there should be a moratorium on such work until guidelines to regulate it could be put in effect. The guidelines that arose from this conference remain in effect today.

In 2015, scientists alarmed about the development of new—and highly precise and easy to use—gene-splicing techniques convened an international meeting in Washington, DC to discuss what to do about it. They recommended a moratorium on research and an Asilomar-like meeting to discuss it.[14] Will it be observed? It looks like the answer is no.

And then there's the matter of unintended consequences. For example, the Federal Aviation Administration proposed to ban the practice of allowing children the airlines refer to as "infants in arms" to travel in their parents' laps. As many parents and others pointed out, the requirement of buying an extra ticket would put many traveling families into cars rather than airplanes. Because car travel is so much more dangerous than air travel, the plan would have resulted in an increase in infant deaths.

Similarly, requirements that cars have seat belts and airbags seem actually to have increased the rate of injuries in auto wrecks, as drivers seemingly rely on the safety equipment, not safe driving. So when policy makers attempt to find the best solution to one problem or another, they must contend with unintended as well as intended consequences.

There are two main approaches to evaluating the merits of action versus inaction or reaction.

One approach is called benefit-cost analysis or BCA. It aims to calculate the benefits and harms of a particular line of action. In its simplest form, a benefit-cost analysis would suggest that if an action's benefits outweigh its harms—by some measure or measures—it's a good plan. If not, don't do it. BCA analysis is widely applied in the United States, especially when advocates of regulation want to prove that its benefits outweigh its costs and opponents want to prove the opposite.

The second approach is called the precautionary principle. Its goal is to be wise before it is too late, which often involves acting even before there is proof of danger, especially if the feared harm might be irreversible. This approach is widely used in Europe; in fact, it is written into the Maastricht Treaty, otherwise known as the Treaty on European Union. In effect, the approach requires anyone seeking to introduce a new product or process to demonstrate that it will not cause problems, now or down the road.

Each approach has its useful features and its flaws, its partisans and its detractors. Critics of benefit-cost analysis accuse its advocates of being apologists for industry. Critics of the precautionary principle accuse its adherents of pandering to populist antiscience sentiment.

Both are highly vulnerable to spin.

As a tool for evaluating the wisdom of regulation and other actions, benefit-cost analysis goes way back. According to James K. Hammitt, who advocates the approach, benefit-cost analysis is more or less what Benjamin Franklin was talking about when he called for "a prudent moral agenda" for policy makers, a principled method to account for and balance competing interests and factors.

Hammitt is a professor of economics at the Harvard School of Public Health and a director of its Center for Risk Analysis. At a conference at Harvard Law School on risk prevention, Hammitt outlined the advantages of the approach.[15] In theory, he said, a benefit-cost analysis looks at action proposed to reduce a particular risk and calculates how much the target risk would be reduced; what other benefits, if any, would result from the action; what countervailing risks or harms would arise because of the action; and what benefits would be eliminated or reduced by taking the action (opportunity costs).

The calculation also notes that benefits and harms from the action might not be distributed equitably—that some people would benefit more than others, and some would face more than their fair share of cost or harm.

As with any model, a benefit-cost analysis can be torqued. And, as Hammitt noted, the term has been hijacked by the right. Opponents of government regulations, especially environmental regulations, have drawn up benefit-cost analyses they say demonstrate that any number of proposed regulations will do more harm than good. "We need to be more modest about what benefit-cost analysis can do," he said in a presentation at the Nieman Foundation at Harvard.[16] Nevertheless, he said, such analyses can help policy makers identify reasonable approaches and areas of uncertainty.

For one thing, he says, the process is transparent. Someone who reads a benefit-cost analysis can see what factors have been included, how much weight they have been given, and what has been left out. The result, ideally, is a framework for comprehensive accounting. In the absence of this kind of outline, he said, we are left to make holistic judgments, something he said we are not too good at. "We tend to emphasize one or two or three highly salient elements" and neglect others that may be important, he said.

Though some call the approach elitist, it can also be populist, if the calculation of benefits and harms takes account of the benefits and harms to everyone, not just to those with advocacy groups or high-priced lobbyists to represent them.

Benefit-cost analysis emphasizes that all choices, even the choice to do nothing, have some benefits and some harms. "Tradeoffs are ubiquitous," as Hammitt put it in a talk at Harvard.[17]

BCA's advantage as an analytical tool is that, in theory at least, a wide range of opinions can be brought to bear on any question. But the approach has flaws, some minor, some major.

In making these BCA calculations, analysts need a way to measure and compare how important risks and potential harms and benefits are. As a practical matter, the results of these calculations are usually expressed in monetary terms. People are misled about this when analysts talk about the worth of reducing the risk of death from, say, exposure to a pollutant. Analysts calculate this figure by asking people how much they would pay to reduce the current mortality rate by one in a million. If, on average, people say that would be worth $5, the value for the proposal overall would be $5 per million.

Opponents of BCA denounce the approach as setting a dollar value on human life and then assuming the use of the calculation implies that someone prepared to pay the money should be allowed to wreak whatever environmental or other havoc they wish. That is not the way the calculation works. It is not a measure of the intrinsic worth of a person, or an average of what people would pay to avoid death. It is an analytical construct, almost a term of art—the value of statistical life, not the value of a life.

But policy makers may not always receive all the information they need to make good decisions. Information they receive may be accompanied by interpretive spin that leads them to ignore or discount or overemphasize important aspects of the problem.

"Everyone knows this is a numbers game," Michael Baram, a member of a panel on radiation hazards in the Carter administration, said as he left the law school conference.[18] When people analyze proposed regulations, they "have to find a bottom line that will not advance the law in a way that is unacceptable to industry . . . what you think industry will accept and what you think your boss will find acceptable." He added, "this is not a monastic enterprise."

BCA also puts a lot of emphasis on probability, and if our estimates of probability are low enough, we may neglect a real problem. Hammitt gave this example, one actually put forward at the time by climate deniers: if there is a 70 percent probability that human actions are altering earth's climate, and there is a 70 percent probability that we could act effectively to reduce the problem, multiplying these two tells us the chances that action will be effective against a real problem (.70 × .70): 49 percent. Conclusion: don't bother. Today, climate scientists, as a group, say the likelihood of human action causing climate change is way more than 70 percent.

BCA should not be used to tell policy makers what to do. It can offer predictions about consequences if this course or that course is chosen, but it cannot say which outcome is more desirable. And, of course, it cannot say how much current generations should sacrifice in order to preserve future generations from harm.

"BCA doesn't grapple very well with that problem," says Joseph Aldy, a professor at the Kennedy School of Government at Harvard who worked on the Clinton administration's analysis of the Kyoto Protocol. Aldy said he and his colleagues did not attempt to do a BCA for that climate accord. It was too complex, he said, and the impacts of climate change were too long term. "If you try to explain benefits one hundred or two hundred years in the future, people have already tuned you out."

So what are the alternatives to BCA?

One might adopt a "sustainability" standard and attempt to meet the needs of the present generation without compromising the prospects for future generations. That might seem like an obvious choice, from the point of view of fairness alone. Unfortunately, there is ample evidence that we are not up to this standard. Whether the issue is climate, water quality, or the use of antibiotics in the livestock industry, industrialized societies have a long record of living in the present and leaving future generations to fend for themselves.

Also, forestalling development can leave future generations with less money and without the progress development can bring. (This argument is a principal prop of the climate deniers—acting against climate change will cripple the economy whereas business as usual will create wealth we can use to engineer our way out of the problem.)

One might also adopt the approach that regulations that don't pay big dividends are not worth considering in the first place. But that would shut us out of a lot of worthwhile actions. If something has a negligible benefit but a very low cost, it might still be worth doing. After all, it *is* worth buckling your seat belt.

The precautionary principle is the major alternative to the BCA approach. It is not a method so much as an outlook on the world. The approach has three premises: first, that the innovative powers of science and technology are increasing so fast we can no longer see where they will take us—they are outrunning our headlights, so to speak; second, that human interference with the environment is now so widespread that there is a growing risk our collective actions can cause serious global effects; and third, that these effects may create unstoppable consequences before we even realize what is happening.

The approach assumes that reducing humanity's impact on the environment is highly desirable. And it requires abundant research on hazards and assiduous monitoring to detect signs of trouble as quickly as possible. It concedes that action to prevent a problem should not be out of proportion to the impact of the problem itself. Perhaps most important, it requires regulatory action at the first suggestion that there may one day be trouble, which may be long before there is any evidence of harm.

Used properly, advocates say, precaution requires that policy makers acknowledge and respect ignorance and uncertainty. They should not focus only on the most straightforward and direct issues but rather realize that there may be indirect impacts of acting or not acting. But of course indirect impacts are hardest to discern and measure—especially if they have yet to occur.

And the good baseline studies and good long-term monitoring of the environment that the approach requires are not exactly magnets for research funds. In fact, the lack of good baseline information has been a major issue in dealing with problems as various as mad cow disease, beach erosion, and the presence of pharmaceutical residues in our drinking water.

Of course, there are objections to the precautionary principle. Taken to its logical conclusion, it would require the inventor of any new product or process to prove it could never be harmful, a requirement that would stifle innovation.

Some politicians and lobbyists view it as a kind of regulatory Trojan horse, an approach that, once adopted, opens the door to all kinds of unwanted regulation. Also, putting it into effect requires a central government with much more power than we give to our checks-and-balances federal government, designed as it is to reduce the chances that sweeping action of any kind will ever take place. Worse, it requires a government willing and able to act before the feared problem has actually materialized.

Also, the kind of latitude for action assumed by advocates of the precautionary principle can allow regulators to act in ways that have little to do with alleviating real risk. For example, the European Union bars the use of genetically modified organisms in human food crops, citing the possibility that they may harm people who consume them. Given that Americans have been gobbling genetically modified foods for years, their health risks seem, at most, unproved.

In this case, however, it is a least arguable that what really motivates the European Union is not concern about health but rather concern about the region's heavily subsidized farmers. The precautionary principle has turned out to be a highly effective tool to keep genetically modified American crops out of European markets. The tool is so effective, in fact, that in 2002, when Zambia and Zimbabwe suffered a famine and the United States offered shipments of corn, the countries turned them down, fearful that if any modified U.S. corn found its way into the countries' seed stocks, they would never again be able to export to Europe.[19]

The fact that there is no requirement for evidence of actual harm means, in effect, that whoever frames the argument wins the argument. In Hammitt's view, for example, George W. Bush's decision to invade Iraq was "classic precautionary principle—we can't wait until we are sure" about weapons of mass destruction.

But advocates of the precautionary principle argue that the BCA method has not served us well either. For example, according to Joel A. Tickner, a professor of environmental health at the University of Massachusetts Lowell, we have some data on hazards of individual chemicals, but hardly any information on what happens when they combine in the environment. We know a bit about what happens when laboratory animals are exposed to large doses, but not much about what happens when people are exposed, chronically, to a suite of artificial chemicals at low doses.

Under the 1976 toxic substances control act (TSCA), he said, industrial chemicals are considered safe until they are proved dangerous—and 99 percent of chemicals now on the market have not been tested. "It's a reactive focus," he said at the Nieman Foundation at Harvard. In his view, unrecognized risks are still risks. Uncertain risks are still risks. Denied risks are still risks.

But removing industrial chemicals from the market until they are proved to be safe, individually and in combination, is not realistic. Given that similar conditions apply with all sorts of products, processes, and procedures that might ultimately turn out to be bad for us, what should we do?

A first step is to acknowledge that there will be problems and uncertainties regardless of which approach we adopt. Interpretations of statistics are not necessarily clear cut. It is not always easy, or even possible, to know which factors are crucial and which can be dismissed. It is difficult to communicate complicated findings in science and engineering to policy makers interested in simple certainties.

Two big new environmental issues face us now and, so far anyway, we seem to be avoiding asking the hard questions.

One relates to nanotechnology, the manufacturing of devices and particles at nano scale—in the billionths of a meter or about one ten-thousandth of the diameter of a human hair. Materials engineered to nano scale have many uses. For example, nano–zinc oxide in sunscreens sinks right into your skin, rather than leaving you with the lifeguard's white nose. But when you take a shower, the stuff ends up down the drain and, eventually, in rivers, streams, and the water supply. Is that a problem? I don't know. But nano-products are moving into the environment by the score.

The second issue relates to proposals to protect us against the effects of greenhouse gases in the atmosphere by indulging in something called geoengineering, the deliberate alteration of Earth's climate. For example, we could inject chemicals like sulfur dioxide into the atmosphere to make it brighter and reflect more of the sun's heat back out into space.

Should we do that? I don't know the answer to that question either. But I believe it is a question for society as a whole to answer, globally. It should not be decided solely by some kind of technical elite, no matter how advanced.

QUESTIONS TO ASK

In their book *Thinking in Time,* Robert E. Neustadt and Ernest R. May offer some terse advice for people facing, as they call it, "a situation prompting action."[20]

First, determine what is actually going on. Identify the problem (and whose problem it is). Then, list what it known, unclear, or presumed about it. Identify analogous past situations and determine the ways they are similar and dissimilar. Only then, list and review options.

After making a list of options, follow these additional steps:

- Acknowledge ignorance and uncertainty. This task may be difficult, because there may be intense disagreement about how much knowledge is enough and how much certainty is necessary. Acknowledge this disagreement.
- Provide money and research support for baseline studies and long-term monitoring. One of the reasons the British government was able to maintain for so long that there was no evidence that mad cow disease (bovine spongiform encephalopathy, or BSE) presented a problem in British herds was that it had never looked very hard for any evidence. Today, the same assertions are made about American herds, with the same lack of evidence. Without independent information, "regulatory appraisal frequently fails due to the dependence of risk assessment on information produced and owned by the very actors whose products are being assessed."[21]
- Ensure that real world conditions are accounted for. The report on BSE by the Harvard Center for Risk Analysis asserted that regulations on animal feed would prevent

the disease from breaking out in American herds. If such standards were adhered to, they probably would. But the report did not take into account the disregard for feed regulations.

- Scrutinize proposed solutions and reject those that only transfer the problem to someone else, the way taller smokestacks reduce local pollution by sending emissions farther away.

- Recognize that once a technology comes into wide use, there will be heavy pressure to continue using it.

- Take wider social values into account.

- Maintain regulatory independence from economic and political special interests. It is hard to imagine how this recommendation might ever be put into effect, absent draconian campaign finance reform, but obviously it is crucial. In Britain, responsibility for dealing with the BSE epidemic originally lay with the agency responsible for the well-being of the cattle industry—an obvious, built-in conflict of interest. In the United States, regulation of agriculture, aviation, and other industries is hampered by dual responsibilities of the agencies involved, which are charged at once with regulating the industry and advancing its commercial prospects.

- Acknowledge that in many cases, decisions to adopt one course or another rely on benefit-cost analyses that are defective by definition: it can be easy to calculate the benefits but difficult—sometimes fiendishly difficult—to calculate the costs. It is not difficult to determine the number of club memberships a proposed golf course will support or the pounds of bacon a hog farm can produce. But the resources that are consumed or destroyed are harder to count. Because "no one knows the value of an acre of wetland to a duck," as one scientist involved in this kind of work once told me, wild ducks are largely left out when someone is trying to assess a proposal to

drain a wetland. Or as Albert Einstein put it, "not everything that counts can be counted." Nowadays, some
people hardly make the attempt.

- Differentiate between questions scientists or engineers
  can answer and questions that are a matter of values.
- Recognize that there may be unanticipated, unintended,
  and unwelcome consequences, especially when the situation involves risks whose impacts cannot be predicted
  with certainty, as when the dangers are hypothetical or
  poorly understood.

Taking Things on Faith

Periodically, researchers discover things—signs of water on Mars,
possible evidence of bacteria in meteorites fallen on Antarctica—that
suggest that there might be life elsewhere in the universe. Years
ago, after one of those discoveries, I asked one of our cosmology
writers if he would do a piece for the newspaper about what it
would mean for theology if extraterrestrial life did turn up. "No!"
he told me. "That's not physics, it's metaphysics!"

I did not argue with him but, for a long time after, I wondered
whether I should have pressed him to take the assignment. Eventually, though, I came to agree with him. When life is discovered
away from Earth, the reactions of theologians will be big news, but
news appropriate for writers who cover religious issues. Religious
issues lie outside the realm of science.

Unfortunately, science and religion have been mixed up a lot in
the United States—more, in fact, than in any other developed
country. People who study the issue attribute it to the fact that religious belief generally is stronger in the United States than in other
developed countries—more people profess belief in a supreme
being, and more attend religious services regularly. Followers of religious orthodoxies, predominantly but not entirely conservative

Christians, play an important part in American politics. When people talk about the Republican Party's "base," often that's who they are talking about.

Surveys suggest that in aggregate we are becoming somewhat less observant. According to the Pew Research Center, 77 percent of Americans claim a religious affiliation, down from 83 percent in 2007. Sixty-three percent say they are "absolutely certain" God exists, down from 71 percent in 2007. Overall, the Pew researchers said, declines in religiosity were most pronounced among millennials. Still, the United States remains by far the most religiously observant of the industrialized countries.[22]

Many believers assert that one must choose between religious faith and science. So it is not a surprise that when the National Science Foundation publishes data on people's knowledge of and attitudes toward science, they repeatedly show that while Americans respect the research enterprise and admire scientists and engineers more than almost anyone else, there is at the same time a strong strain of suspicion about science.[23] Technology is moving too fast, many people say, and taking too little account of moral values.

As science advances, it chips away at the authority of religion, at least as far as the natural world is concerned. It is well established by now, for example, that one's sexual orientation is a biological given, and not a lifestyle choice. We know that storms and earthquakes are not a manifestation of divine wrath but matters of physics and chemistry. People who interpret the Bible literally are correct that its accounts of Creation, cosmology, and the geological history of Earth are incompatible with accepted scientific knowledge.

Many researchers are not concerned about the distance between science and religious belief. But some, especially researchers with strong religious convictions themselves, find the situation tragic, particularly when they meet high school or college students fascinated by science but fearful of losing their faith. But conflict between religion and science is far from inevitable.

The study of science "need not lessen or compromise faith," the evolutionary biologist Francisco Ayala wrote in his book *Darwin's Gift*. "Science can neither prove nor disprove religion."[24]

Unfortunately, many people find it impossible to separate their thinking on science and religion. In the administration of George W. Bush people were named to important government positions because of their religious views, notably their opposition to abortion rights. The administration embraced theories like abstinence education, despite abundant evidence that the programs are ineffective at preventing teen pregnancy. (In 2011, researchers reported what happened when they studied the issue. After controlling for factors like education, socioeconomic status, access to contraception, ethnicity, and other factors, they found "increasing emphasis on abstinence education is positively correlated with teenage pregnancy and birth rates." In fact, they wrote, the policy "may actually be contributing to the high teenage pregnancy rates in the U.S."[25])

Bush is no longer in the White House but religious pressure, particularly from the Christian right, continues to inhibit science in the federal government. Scientists studying issues like gender identity, sexually transmitted diseases, and the ethical treatment of the dying are particularly vulnerable.

But by far the biggest and longest-running public conflict between science and religion involves the teaching of evolution in public schools. Time and again courts have ruled that creationism is a religious idea out of place in a taxpayer-financed classroom. But the controversy does not go away.

Creationist arguments vary. Some adherents are Young Earth creationists—they read the Bible literally as declaring that the universe was created in six days a few thousand years ago. Others concede the universe is much older, but they reject the idea of evolution. Still others accept some parts of the theory, but not others. Among them are advocates of Intelligent Design (ID), the idea that certain features seen in animals on Earth, including people, are too complex to have evolved through Darwinian natural selection. A "designer" must have intervened somehow. Intelligent Design holds, for example, that the eye, the clotting mechanism of the blood, and the tails (flagella) of certain bacteria could not arise from evolution, a more or less incremental process, but had to have been created all at once. Mike Pence supports the idea.

All of these ideas have been disproved again and again. But when ID supporters assert there is legitimate scientific disagreement on this point, few members of the lay public are sufficiently informed about developments in evolutionary biology to know that the idea is, as Ayala puts it, "simply false." So the creationists' argument—"teach the controversy"—continues to resonate.

As an American, I am proud that the idea of giving everyone a say in a controversy carries so much weight. But when it comes to the theory of evolution, there is no controversy to teach. Whatever intelligent design may be, it is not a scientific theory. Its reliance on supernatural action puts it in the realm of the metaphysical. It is a religious idea.

That is not to say the theory of evolution is "true." Nothing in science is "true." One day someone may spot a thread hanging from the theory's metaphorical hem, tug on it, and cause the whole thing to unravel, as happened when Max Planck tugged on the fabric of physics. But the theory of evolution has been subjected to unusually harsh scrutiny since Charles Darwin (and Alfred Russel Wallace) enunciated it more than 150 years ago. So far it has held up. In fact, evidence to support it accumulates by the day. Today, the theory of evolution is by all accounts the most abundantly supported theory in all of science.

In other words, there is no credible scientific challenge to the theory of evolution as an explanation for the diversity and complexity of life on Earth. That was the language I came up with when I first began covering the dispute for the *Times*. I use it routinely. Creationists hate the language. On creationist websites I have seen it referred to as "Cornelia's Creed."[26]

As Ayala put it, though, "there is no controversy in the scientific community about whether evolution has occurred."[27] "On the contrary, the evidence supporting descent with modification, as Charles Darwin termed it, is both overwhelming and compelling." As a result, he concluded, "scientists treat the occurrence of evolution as one of the most securely established of scientific facts."

It is easy to understand why people who believe the Bible offers a literally true account of the creation of the universe may be uneasy about their children being instructed that something else is the case. But parents are not the only ones advancing creationist ideas. The Discovery Institute, a think tank based in Seattle, provides another major source of encouragement.

In 1998, it produced "The Wedge Strategy," a rationale and strategy for "the overthrow of materialism and its cultural legacies." (Note that in this case "materialism" means not greed or a desire to accumulate possessions but rather the central idea of science, that scientists look in nature—the material world—for information about nature, and test that information by experiment and observation there.)

This kind of thinking, the document said, has challenged moral standards, economic ideas, and thinking about society. Eventually, it goes on, "materialism spawned a virulent strain of utopianism. Thinking they could engineer the perfect society through the application of scientific knowledge, materialist reformers advocated coercive government programs that falsely promised to create heaven on earth."

The document laid out a timetable for changing cultural attitudes toward science, starting with research fellowships and moving on to "opinion-maker conferences," teacher training, op-eds, and television appearances, and ending with "cultural confrontation & renewal," more or less urging society to turn its collective back on the scientific enterprise.

In 1999, the plan, described in what became widely known as "the Wedge Document," was posted online. It does not now appear at a Discovery Institute site. I found it in several other places, including the website of the National Center for Science Education, a group that advocates for science-based education. Some people call the organization a "SWAT team" for science.[28]

The document caused something of a stir. The Discovery Institute has described the hoopla as "Darwinist paranoia," and on its website has put up a document entitled "The 'Wedge Document': 'So What?'" The institute denies that it aims to abolish the separation of church and state, and asserts that it is only "as it happens" that so many of its associates see new discoveries in science as "consonant with a 'broadly theistic' world-view."[29]

Anyway, in survey after survey, only about 45 percent of Americans declare that they accept the theory of evolution and about half the population "has significant sympathy for creationism."[30] Surveys find 25 to 30 percent of us are "fundamentalist and resolutely opposed to evolution." The figures have not changed much in the last few decades.

We hear about disagreements over the teaching of evolution when a court case erupts or when state officials attempt to alter curriculum requirements to force the teaching of creationist ideas. But in my opinion the real story of creationism in the United States does not play out in court. It plays out in schools all over the country wherever school boards and superintendents pressure principals and teachers to avoid or gloss over a classroom discussion of evolution for fear of sparking a local uproar. Superintendents or principals may quietly urge teachers to drop the subject or give it short shrift. Also, as I learned in reporting about the teaching of evolution in public schools, officials of the National Science Teachers' Association estimate that about a third of their members embrace creationism in one way or another.[31]

Michael Berkman and Eric Plutzer, political scientists at Penn State, studied attitudes of high school biology teachers on the teaching of evolution. In a 2015 report of their work they referred to the teachers as "enablers of doubt."[32] They said 60 percent of them spent little time teaching the subject, even though evolution is the foundation upon which all of modern biology and medicine are

Pi

From time to time you will see a report that a city council or state legislature or other governing body has redefined the value of pi, the ratio of the circumference of a circle to its diameter. The value of pi is usually given as 3.1415, but it cannot be calculated exactly. Pi is what mathematicians call "irrational"—and the numbers to right of its decimal point run on and on, never ending and never falling into a repeating pattern.

Why would anyone want to redefine it? The answer lies in the story of Solomon who, I Kings 7:23 (King James Version) tells us, built what is variously translated as a vessel or tank with a diameter of 10 cubits and a circumference of 30. That would mean pi=3.

The first I heard of this idea was in an internet link to a site reporting that the Indiana state legislature, citing the

Bible, had declared the value of pi was 3. The report was what we in journalism call "a story too good to check"—a hoax.

Biblical literalists know the Solomon's tank has a pi problem, but they explain it by noting that measuring techniques were not very good in those days. Or maybe the thickness of the tank somehow confused things. Anyway, they add, it's a mistake to take those measurements at face value.

As I read this kind of commentary I wondered briefly why creation literalists could not take a similarly metaphorical view of the story of Genesis. The answer came to me at once. Though the value of pi is mathematically crucial, it has nothing to say about us and our place in the universe. The creation story has everything to say about all that.

based. A further 13 percent hold creationist views or are sympathetic to those who do (other studies put that figure even higher). Only about 28 percent provided what Berkman and Plutzer described as thorough instruction on evolution.

As long as these attitudes and practices persist, it will be hard to change things. Too many students don't learn what they need to know about evolution because too many teachers have not learned it themselves, or have religious scruples against it.

Is It Religion or Science?

Sometimes it is obvious that an assertion or demand is grounded in religion rather than science. But deciding whether something is legally acceptable is a matter for the courts in a country like ours, where separation of church and state is enshrined in our founding documents. The courts typically look to the First Amendment to the Constitution, which says in part, "Congress shall make no law respecting an establishment of religion or prohibiting the free exercise thereof." (The first part is called the Establishment Clause, the second part the Free Exercise Clause.) Those clauses are what courts typically cite in ruling on the issue.

The Supreme Court stepped into this debate in a big way in 1970 in *Lemon v. Kurtzman,* a case involving litigation in Pennsylvania and Rhode Island over financial and other assistance provided by the states to parochial schools.[33] According to the test, a government activity—be it teaching creationism in a public school or setting up a Nativity scene on the City Hall lawn—violates the First Amendment's Establishment Clause if it has the primary purpose of advancing religion, has the primary effect of advancing religion, or creates too close a connection between government and religion.

In another landmark religion case, *Kitzmiller v. Dover Area* [Pennsylvania] *School District,* Judge John E. Jones III relied in part on the *Lemon* test when he ruled against creationists who wanted the school district's science teachers to read to biology students a statement challenging the theory of evolution.[34]

But sometimes the intention is not clear, or its effect is debatable. In that case, courts look to other standards. One of them is the endorsement test, articulated by Justice Sandra Day O'Connor and adopted by a court majority in 1989 in a case involving Christmas and Hanukkah displays outside public buildings in Pittsburgh, Pennsylvania. The endorsement test asks whether a reasonable person would assume the government has sent a message of exclusion to "non-adherents" and a message to adherents that they are, somehow, "favored" members of the political community.[35]

A few years ago, I received a tip that bookstores at the Grand Canyon National Park were selling a book that described Earth as having been created in six days, 6,000 years ago, and added that the Grand Canyon had been caused by a flood sent by God to wipe out "the wickedness of man."[36] (In fact, the universe is about 13.5 billion years old, Earth is about 4 billion years old, and the Colorado River began carving the canyon millions of years ago, and continues to do so today.) The book is *Grand Canyon: A Different View* by Tom Vail who, among other things, leads Bible-oriented tours of the canyon.

"The book purports to be science when it is not," Park Services geologists protested. "The book repudiates science." The presidents of seven geological and paleontological organizations wrote to the Park Service to ask that the book be removed. Years later, it was still on the bookstore shelves.

Going forward, evolution is unlikely to remain the only focus of dispute involving religion and science. Arguments over the age of Earth are an issue, as are challenges to the Big Bang theory of the creation of the universe. Perhaps neuroscientists have the most to worry about. As our understanding of the brain improves, researchers are looking at neurological factors involved in moral choices and even asking what it means, at the cellular level, to speak of something called "the soul."

It is not for the author of a book like this to offer guidance to anyone on matters of religion, except this: it is possible to believe in science, to *be* a scientist, and believe in God—not the God of biblical literalism, perhaps but, as a survey by the journal *Nature* once put it, a God to whom one can pray in expectation of receiving an answer. The evidence for this assertion is among the many scientist-believers. Brother Guy Consolmagno and Father George Coyne, researchers at the Vatican Observatory, an astronomical organization operated by the Holy See, offered an interesting perspective in a dialogue on the National Public Radio program *On Being*.[37] Religious faith begins with observation, they said, and, in a way, scientific research begins with faith—belief that an idea is valid. And, as Brother Consolmagno, the head of the observatory, noted, "the opposite of faith is not doubt. It is certainty."

Just as it is in science.

CONCLUSION

This book has been hard to write. Or rather, it has been hard to stop writing. Too much is happening at the intersection of science and public policy.

As I write, the American Academy of Pediatrics is at war with the Florida state legislature, which has passed a law barring doctors from asking patients about guns in their homes—a known safety risk. Needless to say, the National Rifle Association backs the law, which it says will protect Florida parents from the "political philosophy" of their doctors.[1] The pediatricians' group noted that "at least 10 other states have introduced similar bills . . ."

Also in Florida, the Florida Center for Investigative Reporting discovered that Governor Rick Scott's administration ordered employees, contractors, and volunteers of the state Department of Environmental Protection not to use the terms "climate change" and "global warming" in official communications. One who did was suspended for two days, the *Miami Herald* reported.[2]

In Wyoming, the legislature has banned the collection of information about violations of antipollution laws—like data on water pollution. The legislature says this kind of thing amounts to interference with "economic activity," apparently a reference to ranchers who graze their cattle on public lands, where manure pollutes rivers and streams.[3]

In Congress, lawmakers who regard the Clean Air Act, the Occupational Health and Safety Act, the Consumer Product Safety Act, and similar legislation as needless and costly interference with corporate America are proposing a raft of laws that, in effect, turn scientific decision-making over to political appointees or even politicians, rather than the government scientists who did the research. As a group of outraged researchers expressed in an essay in the journal *Science,* "congressional leaders are pressured to render these long-standing and well-regarded laws ineffective by undermining their scientific foundations."[4]

All of these policy-making disasters are distressing to contemplate. But these issues are comparatively straightforward, compared to the far knottier scientific and ethical questions we will face soon. For example, in January 2015, a drone attack on an Al Qaeda stronghold in Pakistan resulted in the deaths of two hostages, one of them American, the other Italian. Tremendous uproar ensued. Imagine public reaction the first time a U.S. hostage is killed by friendly fire, and the attacker is not a drone controlled by a soldier at an Air Force base in Nevada, as many U.S. drones are, but rather is an autonomous battlefield robot making its own decisions about when to fire, and at whom. Such robots are under development.

Improvements in sensor technology, advances in artificial intelligence, and other developments will soon make it possible to augment drones with "weapons systems that make their own killing decisions."[5] Engineers and engineering ethicists say that might actually be a good thing, in that an autonomous battlefield robot will not be angry, cold, hungry, grief-stricken over the death of a buddy, or otherwise prone to acting out.

On the other hand, this kind of technology can malfunction or be hacked or even (possibly) elude control, says Paul Scharre, a theorist on the future of warfare at the Center for a New American Security.[6]

Is this a technology that should come into use without a public debate?

As the greenhouse effect continues to warm the planet, support grows for the idea that we should deliberately modify Earth's climate through the use of geoengineering techniques. The debate about these techniques is already moving from engineering academe to the real world.

Several groups have tried altering ocean chemistry by dumping iron filings into the sea in order to encourage the growth of tiny ocean plants called plankton. In theory, dumping iron in iron-poor ocean regions would encourage the plants' growth. They would absorb CO_2, removing it from the atmosphere. Eventually they would be eaten by fish or other marine creatures, or die and fall to the ocean bottom. After a private company's unsuccessful attempt in the Southern Ocean, the United Nations declared that there should be no more of this, at least for now. Despite this call, in 2012 researchers working with the Haida First Nation of British Columbia applied 120 tons of iron to waters west of Haida Gwai.[7]

Iron fertilization is not the only kind of climate-altering experimentation that people are thinking of. Other techniques, such as injecting chemicals into the atmosphere or inserting mirrors in stationary orbit between Earth and the Sun, will have greater effects.

Of course, all of us who use fossil fuels have in effect been conducting an inadvertent experiment on Earth's climate since the beginning of the Industrial Revolution. But decisions about going ahead with *deliberate* climate-altering experiments should not be made by individual groups or companies or even nations. Their potential implications are too vast. So far, though, there has been little discussion of who should be in charge of this work, who should decide when to apply it, and who should determine what ill effects would be acceptable, much less who would have to accept these ill effects or how those people will be compensated, and by whom.

In several labs around the country, researchers are working to enhance the killing power and spreadability of disease-causing microbes like flu, virus relatives of smallpox, and other pathogens. When I first heard about this work, I could not believe that any right-thinking person would undertake it because its safety risks and potential for misuse are so obvious. But its advocates claim it could help in the development of vaccines or help identify emerging epidemics.

The work reminds me of another, quasi-related issue I wrote about in 2006, amid widespread worry about new strains of flu. In particular, the issue was the use of breathing machines called ventilators. The machines can keep sick people going until they recover enough to breathe on their own. But in the event of a pandemic, there won't be enough machines or people trained to use them. How will we decide who gets them, and who does not? What if we have to take some people off ventilators, condemning them to death, so that other patients might live? I was prompted to write about the issue because of a report by New York state health officials who argued forcefully that decisions on these matters should not be decided by technical elite. Rather, they stated, these are questions of social values, which should be widely discussed so that the policy that results has wide social acceptance.[8]

By now, a number of other states, including California and Maryland, the Department of Veterans Affairs and a few other agencies have begun working on what amount to protocols for rationing. But their discussions are taking place largely inside the halls of science, with participants who sometimes have powerful conflicts of interest.

What may be the most significant development of all involves the development of a simple, cheap, and powerful genome engineering method that could be used not just to repair the genes of a person with an inherited condition but also to alter people's DNA such that the repair will be passed down to their children and their children's children, indefinitely.[9] The change seems to herald an era in which designer babies are real possibilities. Humans will have the power to alter the species, permanently.

In a commentary in the journal *Nature,* one of the developers of the gene-editing technology, Jennifer Doudna of the University of California, Berkeley, pleaded for greater, wider public discussion of the ethical issues the technology raises. The new method, called CRISPR-Cas9, "has changed the landscape of biology," she wrote.[10] Doudna said that as public discussion of the technology proceeds—and she assumes there will be such a discussion, though she does not define "public"—five steps are necessary:

- researchers need to adopt standard methods and metrics to assess their experiments;
- scientists need to provide information to the public "about the scientific, ethical, social and legal implications";
- policymakers and scientists should collaborate on determining what is (or is not) an ethically acceptable use of the technology;
- "appropriate oversight" arising out of these discussions should be organized and applied; and
- work should not proceed on alterations to the human germline because the technology is not yet fully developed—but also because of the "unknown social consequences" of the work.

In short, the ethical, political, legal, and other social issues relating to this work are at least as important as its technical issues, and they should be discussed outside the walls of the world's labs.

Gene editing is one of several new biotechnologies that could remake our relationship with our species, and other species.

In May 2016, nearly 150 researchers working with DNA attended a closed-door meeting at Harvard Medical School to discuss the possible creation in the lab of DNA—and of human chromosomes. Organizers of the meeting declared that its purpose was aimed at synthesizing DNA in general and not the creation of "human beings without biological parents."[11] But the gene scientist J. Craig Venter has already created a bacterium from scratch. It's hard to rule out the eventual synthesizing of a human genome.

As the *New York Times* put it, in a triumph of understatement, embarking on such a project "would raise numerous ethical issues." A few weeks later, the National Academies of Sciences, Engineering, and Medicine endorsed research on a technology known as "gene drive," which could in theory be used to alter the genes of entire species—or even eliminate entire species.[12] Such a technology offers many attractive possibilities, such as the elimination of disease-causing parasites or even disease-transmitting organisms like the mosquito or the deer tick. (Researchers on Martha's Vineyard, which is heavily afflicted with tickborne illness, are already looking at the technology.) But the possibilities of unintended consequences are obvious.

At the end of 2015, representatives of the National Academies, the Royal Society (the U.K. equivalent), and the Chinese Academy of Sciences met in Washington, D.C. to discuss how to confront this new science. Among other things, they called for wider discussion of the technology among a wide swath of people including researchers, public officials, religious, ethical, and legal experts—and ordinary people.

Many researchers concerned about these issues fear that organized discussion of this kind of technology is likely to follow what is now called "the Asilomar model," emphasizing "scientific and technical expertise over expertise in governance, risk management, and organizational behavior. Political leaders have largely ceded a strategic leadership role," wrote Megan J. Palmer, a Stanford expert on biological engineering, and her colleagues in the journal *Science* after the Washington meeting.[13]

Given what we have seen with political debates in this country over climate, gun control, contraception, and other issues, it may be that giving scientists a decision-making role is better than the alternative. But that is a defect in our politics, not any kind of proof that political leaders *should* be kept out of discussions of this kind.[14]

I hope that the research community embraces the necessity of bringing members of the public into these discussions. Daniel Sarewitz, the co-director of the Consortium for Science, Policy, and Outcome at Arizona State University, Tempe, summarized the situation in a commentary in the journal *Nature* in 2015. He wrote: "The idea that the risks, benefits and ethical challenges of these emerging technologies are something to be decided by experts is wrong-headed, futile and self-defeating. It misunderstands the role of science in public discussions about technological risk. It seriously underestimates the democratic sources of science's vitality and the capacities of democratic deliberation. And it will further delegitimize and politicize science in modern societies."[15]

As he noted, continuing nonsensical "debates" over all kinds of issues—vaccine safety, GMO crops, and cell phone safety among them—"should be evidence enough that science does not limit or resolve controversies about risk." On the other hand, he added, "the capacity of people to learn about and deliberate wisely on the technical aspects of complex dilemmas has been documented by social scientists for decades."[16]

As I write, the United Kingdom has just given a researcher permission to alter human embryos using Crispr-Cas9 gene-splicing technology. The researcher says she will allow the embryos to die and will not implant them in a womb. We'll see what happens next.

Progress in science and engineering is always going to be ahead of progress on the philosophy or ethics of using the technology. But the moral and philosophical issues are by far the more important.

Harvey V. Fineberg, a former provost of Harvard and president of the Institute of Medicine, the National Academy's medical arm (now the National Academy of Medicine), addressed this issue in an editorial in the journal *Science* in February 2015.[17] His immediate topic was gain-of-function research (a study in virology of pathogens that are manipulated to alter their capabilities), but his words apply more widely.

"Judgments by those from different scientific and lay perspectives will be critical to sound decision-making," he wrote. "The input of scientists and funders is no longer sufficient to make appropriate, socially defensible choices about research that has such social dimensions."

We ordinary members of the public must step up and demand a place at the table. If not, we may find ourselves in a brave new world we do not like. Or we will see useful technologies and inventions increasingly kept from public use by the influence of technically illiterate scare-mongers or deceptive people who know how to dominate public discussions and distort them with scientific nonsense.

Meanwhile, there are a few things we can do.

- We can acknowledge ignorance and uncertainty. As I have noted, task will be complicated by disagreements, possibly intense, about how much knowledge is enough and how much certainty is necessary. We can acknowledge the disagreement. (As I once told someone who complained that one of our articles would leave readers confused, "if a situation is confusing, confusion is an appropriate response!")
- When it will be useful, which is often, we can support long-term monitoring and other mundane but crucial efforts to gather the baseline data and other needed information before decisions are made. Once we embark on a plan of action, we can demand that results be tracked and that revisions to the plan be made as needed—in other words, we can demand adaptive management.
- We can scrutinize proposed solutions and reject those that only transfer the problem to someone else, the way taller smokestacks reduce local pollution by sending emissions downwind.
- We can recognize that once a technology comes into wide use there will be heavy pressure to continue using it. The difficulty here is differentiating between useful caution and New Age Luddism that rejects technological innovation out of hand.
- We can act to maintain regulatory independence from economic and political special interests. We can do away with the so-called "revolving door" that sees people from a given regulatory agency moving into high-paid jobs in the industry they have been regulating—everything

from agriculture to drug manufacturing and investment banking.[18]

- We can acknowledge that in many cases decisions to adopt one course or another rely on benefit-cost analyses that are defective by definition, in that it may be easy to calculate the benefits but difficult to know the costs.

- We can consider other views by, for example, considering news sources outside of our individual comfort zones. (Note that all news sources are not equally valid: research demonstrates that regular viewers of Fox News are significantly more misinformed than Americans generally.)[19] We can guard against fake news.

- And when you come to a conclusion, consider it again. Know as much as you can about the available options and the science or engineering behind them. Understand not just why you like your own choice, but also what makes it superior to other options.

It seems naive to urge this kind of approach after the political dumpster fire of the 2016 presidential election. But it is our best way forward.

TRUSTWORTHY,
UNTRUSTWORTHY,
OR IRRELEVANT?

In 1979 there was a partial meltdown at a nuclear plant at Three Mile Island, in Dauphin County, Pennsylvania. I was a young newspaper editor at the time, and I was caught up in coverage of the resulting debate about whether nuclear power could ever be safe. I have long forgotten the details of that episode, except for one troubling thought that occurred to me in the middle of it: the experts we relied on to tell us whether a given design was safe, or indeed whether nuclear power generally was safe, were people with advanced degrees in nuclear engineering and experience running nuclear plants. That is, we were relying on people who made their living from nuclear power to tell us if nuclear power was safe. If they started saying out loud that anything about the nuclear enterprise was iffy, they risked putting themselves out of business.

I mention this not because I think the engineers lied to the public. I don't. Nor do I think nuclear power is so dangerous it should be rejected as an energy source. I mention it because it shows how hard it can be to make sense of information from experts.

Here's another example: men with prostate cancer are often asked to choose between surgical treatment and radiation. Quite often, they find the surgeons they consult recommend an operation and the radiologists suggest radiological treatment. This real-life example of the old adage "don't ask the barber if you need a haircut" is just one of the reasons why dealing with prostate cancer can be so difficult.

And finally, the bedbug. In recent years there has been, it is alleged, an epidemic of bedbug infestations in New York City apartments. They have even been reported in offices and theaters. But the evidence comes largely from companies you can hire to deploy bedbug-sniffing dogs that react when the creatures are around. Other pest control experts often found no evidence of infestation.[1] The bedbug folks attributed false positives to poor training of the dogs or their handlers, or the possibility that the dogs were picking up scents brought in on clothing or linens or wafting through ventilation systems from other apartments. Who knows who is right? One thing we do know—it is in the companies' interest for people to believe bedbugs are on the march in Manhattan.

Having said that, I must add that it is unwise to reject information solely because the person offering it has a financial stake in it. This approach has provided a useful weapon for the disingenuous people who want to discredit one idea or another but find the facts are against them. They argue that the experts are speaking from positions of vested interest.

Climate change deniers put this canard forward routinely in their arguments against action on global warming, describing climate research generally as a closed shop whose members somehow profit from a flow of research funds from government agencies that have drunk their particular brand of Kool-Aid.

Creationists raise it in connection with the teaching of evolution— biology as an institution, they argue, has some kind of vested interest in evolution and will quash anyone who challenges it. I doubt it. Most scientists I know would love to be the one who pulls on the thread that causes climate worries to unravel, or frays the warp and woof of biology—assuming such threads exist in the real world.

Anyway, conflict of interest is just one of many factors to weigh when you are considering opinions, supposedly expert, on one issue or another.

Who Is an Expert?

One of the first things to consider is, who is making the claim? Are they people we should take seriously? In other words, are they experts? If you don't at least try to answer this question you may waste time or even delude yourself. As the sociologists of science Harry Collins and Robert Evans put it in their analysis of expertise, "other things being equal, we ought to prefer the judgments of those who 'know what they are talking about.'"[2]

For example, at the Institute for Advanced Study, Freeman Dyson researched solid state physics and related fields. Today, more people know him as a critic of climate change theories. Among other things, he asserts that ecological conditions on Earth are getting better, not worse, an assessment many people would contest.[3] Dyson has a distinguished record in physics, but he is not doing the kind of work that would put him in the front rank of researchers on climate or the environment generally. As one commentator put it, "Many of Dyson's facts on global warming are wrong."[4]

Plenty of other so-called experts have won fame by speaking out on subjects outside their fields. The occurrence I remember most vividly involved a story about the medical researcher Judah Folkman and his theory that a way to thwart cancerous tumors would be to prevent the growth of blood vessels they need to survive. The growth process is called angiogenesis, and Folkman was working on antiangiogenesis drugs. One of our reporters heard about the work, which was exciting a lot of interest among cancer researchers and other scientists. The result was an article that ran on Page One.[5]

The story was very carefully written. Folkman himself was not making great claims—as he said in the article, "The only thing we can tell you is if you have cancer, and you are a mouse, we can take very good care of you."

But someone else we quoted was not so reticent. He was James Watson, the biologist who with Francis Crick elucidated the structure of DNA in 1953. "Judah Folkman will cure cancer in two years," Watson declared. That really caught people's attention. But tumor growth was not Watson's field. He was not someone we should have sought out for comment on the work. And of course he was wrong—though anti-angiogenesis drugs have found wide use against several cancers, cancer remains unvanquished. Today, when people ask me if there is any story I regretted doing at the *Times* I can answer truthfully that there is not. But I wish we had not cited Watson as an expert in that story.

But who is an expert? Massimo Pigliucci, a philosopher at the City University of New York, puts the question this way: "How is the average intelligent person (Socrates' 'wise man') supposed to be able to distinguish between science and pseudoscience without becoming an expert in both?"[6]

Even researchers struggle for ways to define leadership in their fields. Over the years, a number of metrics have come into use, but all of them have deficiencies, and they are not the kind of information the average person has (or necessarily wants). Still, you might want to know about at least a few of them, if only to get a sense of how difficult it is to judge. Usually, these measures of scholarly leadership relate to the degree to which other researchers cite a scientist's work—"a systematic effort to track citations—the footnotes by which journal authors acknowledge their intellectual debts."[7]

The first of these metrics is the number of times a researcher or research paper is cited by other researchers. The problem here is that those who invent useful research techniques may find themselves cited repeatedly, even if they do not use the techniques to make stunning findings themselves. That is, their achievement is one of tool-making, not discovery-making. Tool-making is not trivial—the development of a way to culture kidney cells as a growing medium was key to the development of the polio vaccine, for example—but it does not alter our fundamental understanding of things.

Another such metric is the "impact factor" of the journal itself—the frequency with which articles in the journal are cited subsequently. This factor speaks to the quality of the journal, though, and not necessarily to the quality of any particular paper in it. A version of this measure called "evaluative informatics" gives greater weight to citations from papers that are themselves widely cited.

A final metric deserves mention here: the *h*-index. This measure, introduced in 2005, gives a researcher with fifty articles, each cited fifty times, an *h*-index of fifty. More recent variants give added weight to newer or more widely cited articles.[8]

Most scientific papers have multiple authors, however, and these methods do not necessarily tease out who was the intellectual engine of the report and who is merely tagging along in the lab.

What does all this mean to you? Only this: when you are trying to make sense of new scientific or technical claims, especially if they fall well outside the mainstream of accepted knowledge, you should consider whether the person making the claim has a track record of findings in the field that hold up, or that fellow researchers consider important and reliable. If you cannot make this judgment, just file the question away in your mental folder of things you'd prefer to know before making up your mind.

Does the person have any credentials? It's good to know. But credentials can be deceiving. I wrote about such a case when I reported on Marcus Ross, a Young Earth creationist who earned a doctorate in paleontology from the University of Rhode Island.[9] Ross, who said he believed that Earth was only a few thousand years old, wrote his thesis on a long-extinct creature that had lived millions of years ago. He reconciled the obvious discrepancy between his religious beliefs and his scientific work ("impeccable," his notoriously demanding thesis advisor called it) by viewing them as representing different paradigms. But even before he earned his degree, he was featured on DVDs identifying himself (truthfully) as a Young Earth Creationist trained in paleontology at URI. As Pigliucci notes, "it is always possible to find academics with sound credentials who will espouse all sorts of bizarre ideas, often with genuine conviction."[10]

Are the expert's credentials relevant? Often people with legitimate credentials in one field begin opining about matters in other fields, in which their expertise may not be particularly helpful. As Pigliucci notes, "It is unfortunately not uncommon for people who win the Nobel (in whatever area, but especially in the sciences) to begin to pontificate about all sorts of subjects about which they know next to nothing."[11] This was our problem with James Watson.

Next, ask yourself where the expert fits in the field being discussed? Is this expert an outlier? That does not by itself invalidate anyone's opinion. The history of science is full of outliers who came to be triumphantly proved right. Usually, though, outliers are wrong.

What biases or conflicts of interest does the expert bring to the table? As we have seen, it can be hard to find out and, once you know, hard to calculate how big an effect, if any, these biases might have, but the issue should be on your mind when you consider whether to trust someone's expertise.

Perhaps your supposed expert is actually a crank. When I was science editor of the *New York Times,* I often heard from people who wrote to say that they had disproved the germ theory or the atomic theory or some other gold-standard landmark of science. I dismissed them as cranks. And I worried every time that I was consigning a brilliant insight to the dustbin. Still, there are a few signals that someone may be, shall we say, somewhat loosely wrapped. Approach with caution those who claim that they do battle against the entire scientific establishment, or that enemies of some kind conspire to deprive them of the recognition they deserve.

Show Me the Data

When you are looking at data, see if you can trace them to their original source. Often, they will turn out to have been "detached from any caveats, or averaged or manipulated in appropriate ways," notes Jonathan G. Koomey, a Stanford professor whose work involves, among other things, critical thinking and information technology. The result? "Mutant statistics." According to Koomey, "Even renowned institutions can publish nonsense, and respected authorities are sometimes way off the mark."[12]

How can this happen? Here's an example, related to me by a participant, its facts changed to protect the guilty. It involves a member of Congress, a press secretary, and the Congressional Research Service.

When the Congressman made a speech, often it was the press secretary who wrote it. One day, he was scheduled to give a speech in Seattle. He was going to talk about timber. In particular, he wanted to talk about spruce, and he wanted to mention how many board feet of spruce are exported each year from the United States.

The press secretary got to work. He wrote a fine speech. But try as he might, he could not find out how many board feet of spruce are exported each year from the United States. He even called the Congressional Research Service, ordinarily a source of abundant and reliable information, and the people there could not help him.

Finally, the Congressman was getting ready to leave to give his speech. The press secretary gave him the text. It said that the United States exports 10 million board feet of spruce each year, a figure the press secretary had made up.

The next week, the Congressman returned to the Capitol. The speech had had gone over so well, he said, that he wanted to enter it in the Congressional Record. But the press secretary was reluctant to enter pure fiction into the Congressional Record. So again he called on the Congressional Research Service. Couldn't they come up with something?

Eventually they did. They called back and told him the United States exports 10 million board feet of spruce each year.

The press secretary was amazed by the astonishing coincidence. But almost immediately he had another thought. "Where did you get that information?" he demanded. "Oh," he was told, "from a speech by Congressman X in Seattle last week."

Moral: it's always good to ask.

We should particularly ask that question when the information involves something secret (proprietary) or when it relates to an axe someone is attempting to grind. As a young reporter in Rhode Island, for example, I learned to discount the numbers when police departments offered the "street value" of drugs they had seized in a raid. Almost always the figures were wildly inflated.

How to Read a Research Study

A friend of mine suffers from vertigo. A while ago, she heard about a new treatment for the malady that involved injecting a drug through the eardrum, and she found a study assessing the value of the treatment. It looked interesting to her, so she brought it to her doctor and asked if the treatment was something she should consider.

She emailed me about what happened next. Her doctor, she said, dismissed the study as "not very well designed."

"I have no idea how to tell if a study is well designed," my friend wrote me. "I can ask him of course. But I thought I'd ask you first."

Well, I am not an expert either. But I could point to a few things that might have been on her doctor's mind. The drug went to people who chose the treatment, rather than being administered randomly. The study was retrospective; that is, it relied on people's memories of their treatment and its results, rather than observation on the spot. Its end point was something the researchers called "survivability," which they defined as the ability to live with the condition the drug left them in. That is a surrogate end point, and a highly subjective one at that.

Does this mean the study would lead my friend in a wrong direction? Not necessarily. But those factors are worth considering.

There are other factors you should consider when you assess a research study. Some journals are leaders in the field—among them are *Science,* published by the American Association for the Advancement of Science (nonprofit), and *Nature,* published in Britain by the Nature Publishing Group (for profit). *Cell* is a leader in biology. In medicine, leaders are the *New England Journal of Medicine* (known as *NEJ*), published by the Massachusetts Medical Society, the *Journal of the American Medical Association* (known as *JAMA*), the *Annals of Internal Medicine,* and two British journals, the *British Medical Journal* (*BMJ*) and the *Lancet.* I don't mean to suggest you can bank on everything that appears in those journals, or that other publications are less reliable. But it pays to consider where research appears.

Let's assume you have determined there is a good reason for you to pay attention to a new research report. How can you make sense of it? Research papers vary, and journals vary in the way they report research, but here is a brief tour through the typical research paper.

Title. Don't worry if it sounds strange, incomprehensible, or even funny. Often they do. That's why it can be easy for the unscrupulous or the ignorant to poke fun at research as a ridiculous waste of money.

Authors. Look (usually in a footnote) for information about the author to whom correspondence should be addressed. Often that person is the so-called "lead author," the person who headed the research. Note, though, that often people are listed as co-authors who had little or nothing to do with the work being described. Perhaps they lead the lab where someone else did the work, or developed a testing technique used by the people who did it. It may be impossible for you to determine who did what to whom in any particular research project, but be aware that the appearance of the names of prominent researchers on a list of a paper's co-authors does not necessarily imply that they would sign off on every bit of it.

David Baltimore, Nobel Laureate and later president of CalTech, learned this lesson the hard way when he agreed to have his name appear on a 1986 report on immunology research of a young colleague at MIT. The work became the center of a drawn-out misconduct investigation. The researcher was eventually cleared. But, in a letter to the National Institutes of Health, Dr. Baltimore described himself as doing "too little to seek independent verification" of the data in the paper.[13]

In part because of that case, some journals now require papers to describe each co-author's contribution to the work. For example, editors of *Science, Nature,* and the *Proceedings of the National Academy of Sciences* have agreed that authors will be required to identify their contributions, and that the senior author must "confirm that he or she has personally reviewed the original data" to ascertain that it has been "appropriately presented."[14]

Abstract. This section, typically a fat paragraph, summarizes what the research is about, how it was done, what it found, and what the findings mean. Reading the abstract can often, though not always, tell you whether it will be worth your time to read the study itself.

The body of the study is typically organized in sections like these:

Introduction. This section elaborates on the question the study seeks to answer, and why.

Methods. This section describes how the study was conducted—whether in a laboratory, in the field, with a group of patients who met particular criteria; how long it lasted; the size of the sample studied; whether test subjects were selected randomly; how the researchers controlled for potentially confounding variables (like age, for example, in a study of whether a given substance is linked to a given disease); how long the study lasted; and the statistical techniques used to analyze the data.

Look particularly at how research subjects were selected. Recently, the Women's Health Initiative overturned decades of medical practice, and challenged results from the long-running Nurses' Health Study, an epic achievement of public health, when it found that women who took hormone replacement after menopause were more likely to have heart trouble, not less.

But as critics gradually began to notice, the study enrolled women whose mean age was 63. In other words, they had been a decade or more past menopause. What the study demonstrated, it seemed, was that starting replacement therapy ten years after menopause offered little benefit.

The results "generated debate and confusion," researchers wrote in the journal *Science*. "(A)spects of the population included in the trial, as well as the hormone regimen, must be addressed," they wrote.[15] In short, the findings may not be generalizable to women who began replacement therapy at menopause and continued it. Research to answer that question is underway.

This section is one of the places where researchers will point to methodological limitations of their work—be they funding limits, restrictions on research involving human subjects, access to remote field sites, or other problems. Here is where they will (or should) acknowledge any effects these limitations may have on their results.

I confess that as a lay reader, even a reader who has spent years looking at research reports, this section often defeats me. Still, it is worth looking at.

Results. Here is where the researchers describe their findings. Unless you have some knowledge of statistics, this section is another opportunity for confusion. What you want to know, though, is whether the findings are statistically significant—defined by scientific convention as whether the probability of their occurring by chance is 5 percent or less, which is usually reported as a "p-value." Look for a p value of 5 percent or less ($p < 0.05$).

Remember that this standard is arbitrary. Even if results are not statistically significant, they may still have some value, and the authors may argue the point. On the other hand, the fact that a finding is statistically significant does not guarantee that it is important, or that it did not occur by chance.

Also look for a discussion of what researchers call the confidence interval. Typically, researchers will report a study result of, say, X and then say they are 95 percent confident the result lies between $X–Y$ and $X+Z$. "Notes from the Science Lab," a website for editors, gave the example of a study in which researchers found that drinking chlorinated water was associated with a 38 percent increase in bladder cancer.[16] But the confidence interval was 1.01 to 1.87. "That meant that while the researchers thought it was 38 percent, it could be anywhere from 1 percent to 87 percent," the site explained. "Obviously, this is a highly speculative study."[17]

"The p-value was never intended to be a substitute for scientific reasoning," Ron Wasserstein, executive director of the American Statistical Association, said in a statement. Over-reliance on it, the association said, can mean that potentially useful research that does not achieve formal statistical significance can end up ignored.

Discussion. This section is where the researchers attempt to put their new findings into the wider context of the field and draw conclusions about their work.

Here is where they may also present alternate explanations for their data. It would be unusual (I would almost say suspicious) for researchers to conclude that their work is the final word on their subject. Far more often, they emphasize issues that remain murky and describe new avenues for inquiry. As a journalist, I am chronically frustrated by this focus on new questions rather than results. But this kind of thing, along with calls for other researchers to test their findings, mark responsible research.

Unfortunately, an emphasis on questions and caveats may create the impression they lack confidence in their work.

References. Here is where researchers cite other work in the field they have consulted or relied on in the course of their research. If you are up on the field, you may be able to tell if something that ought to be here—a signal study or the like—is missing.

Financing. Some journals now require scientists to say who paid for their research. Be aware of where the money came from—or if that information is missing.

Finally, if the scientific report is published in *Science, Nature,* the *New England Journal of Medicine,* the *Journal of the American Medical Association*, or other major journals, it may be accompanied by a news article or editorial. At the *Times,* when we evaluate the importance of one finding or another, we regard this kind of attention as an indication that the journal editors, anyway, regarded the finding as worthy of unusual note. (Also, lay language editorials or news articles, even in the not-particularly-lay-language of scholarly publications, can be helpful guides to the study itself.)

How to Assess Findings

Now let's consider the findings themselves. Here are some features to look for.

- What kind of study was it? Remember the differences between observational studies and controlled experiments. Was the study prospective (designed before data were collected) or retrospective (looking back at previously collected or merely remembered data)?
- Where does the study fit in the field? Usually researchers will report whether their work fits in with previous work, overturns widely held assumptions, and so on.
- Is the finding credible? For example, if the study suggests that A causes B, what is the mechanism the researchers suggest brings all of this about? Is it credible, or even scientifically possible? That issue was one of the signal weaknesses of the mercury-vaccine argument and the cell phone–brain cancer link. Note, though, that if no mechanism is identified, it does not necessarily mean that none exists.
- Does the effect increase as the dose or exposure increases? Researchers call this the dose response. Note that it may not appear if there is a threshold that must be met before any effects are seen.
- Does the study have methodological flaws? What cautionary notes do the authors of the study offer? Responsible researchers will draw attention to any factors that limit the applicability of their findings or weaken their conclusions.
- Was the study blinded? That is, did researchers evaluating results know who was subjected to the drug, treatment, or other intervention and who was not? Unless they were in the dark about it—unless the study was blinded—their assessments may be biased by their hopes for what the research would show.

- What are the compliance and drop-out rates? For example, a drug may improve health for people who take it, but if it has so many evil side effects that people fail to take it properly or abandon the trial altogether, it can tell you something about the drug's usefulness. Differential rates of dropping out can have big effects on study findings.
- To whom or what do these results apply? Who or what is excluded by study design, methodological flaws, and so on?
- How do these results fit into what we know generally about the subject? If its results are valid, would the work confirm earlier findings or throw them into doubt?
- What do other researchers in the field have to say about it? In the Science Department of the *Times,* we always took note when researchers praised the work of rivals or scientists whose theories ran counter to their own.
- Could the results be interpreted another way? In other words, could an intelligent person of good will look at this data and come to another conclusion? Look for the researchers' rationale for interpreting their findings the way they do.
- Is the study generalizable to other groups, places, or subjects?
- Is the study size sufficient to draw conclusions? There may be no way you can independently tell. With medical studies, some experts advise caution if there are fewer than 100 subjects, but that may be arbitrary. Still, a lot of angst might have been avoided if people citing the original (now discredited) *Lancet* vaccine-autism study had noticed how few children were involved.
- Was the study designed to achieve its stated purpose?

As researchers noted in a special report in the *New England Journal of Medicine,* clinical trials are expensive, so "investigators frequently use analyses of subgroups of study participants to extract as much information as possible." A few years ago, for example, researchers studying a candidate AIDS vaccine reported it had no benefit in their test subjects. Possibly seeking some sort of good news in this discouraging result, they pulled out findings for members of minority groups, some of whom appeared to benefit.

Was the result real? That study does not say so. It merely suggests a new line of research. As the authors noted, analyzing a subgroup, rather than the study group as a whole, can "introduce analytic challenges and can lead to overstated and misleading results."[18] When properly reported, the authors went on, subgroup analysis "can provide valuable results." Scientists often disparage this kind of analysis as "data dredging." Researchers should be transparent about how the groups were chosen and how the findings were analyzed, and they should indicate what steps they took to avoid erroneous conclusions.

On the other hand, be aware that sometimes factors that differentiate one subgroup from another are obviously important. David Kent and Rodney Hayward made this point in an article in *American Scientist* in 2007.[19] They offered the hypothetical example of two heart attack patients, one a 72-year-old diabetic whose electrocardiogram indicates a massive heart attack, the other a 52-year-old with no other health issues whose electrocardiogram indicates a minor heart problem. Under current practice, both would probably be treated with clot-busting drugs. But the older, sicker patient is far more likely to benefit from the treatment. The younger patient is far more likely to suffer their side effects without any benefit.

You may think you've thoroughly examined and assessed your research study by this point, but consider a few more crucial questions.

- Does the study adequately control for the impact of confounding factors (age in a health study, say, or weather patterns in a study of beach erosion)?
- Has the research been peer-reviewed?

- Did the study rely on surrogate end points, like cancer diagnoses, or intermediate end points, like tumor shrinkage, rather than end points of interest, like cancer deaths? If so, do you consider its conclusions worth your attention? If the researchers used a surrogate or intermediate end point, how well does it represent the reality you care about?

- Were the findings observed in laboratory dishes (*in vitro*) or in laboratory animals or in actual people (*in vivo*)? Bleach kills the AIDS virus in a lab dish, but it is not a useful treatment against the virus. And while animal studies offer many valuable clues, animals are not small people. Women are not the same as men, and old people are not the same as young people.

- What was the study designed to test? Richard Peto, the Oxford University meta-analyst, once analyzed the effectiveness of aspirin as a treatment for heart attack and found it did not help people born under the astrological signs of Libra or Gemini. Is this a real effect? Undoubtedly it is not. The study was not designed to determine whether astrology played a role in heart therapy. Peto once also analyzed fertility data and found that patients with *g, y,* or *n* in their last names were more likely to conceive.

- Who did the study? Who paid them to do it?

- Who is the study for? That is, who is expected to act on the information—other researchers, clinicians, members of the public, regulators? And what are they supposed to do about it—open a new avenue of research, alter their treatment decisions, change their habits, alter regulations?

- How significant in the real world is the observed effect?

- If the study is a meta-analysis—a study of studies, so to speak—how were the individual studies selected? Are the criteria reasonable?

How to Assess Risk

If the issue is a matter of risk, different questions arise.

- How large is the supposed risk? Is it big enough to warrant changes in behavior or policy? It's worth not smoking, because smoking doubles the risk of heart disease, already a very common problem. But it might be worth tolerating something that doubles your risk of some obscure ailment from 1 in a million to 2 in a million.
- On the other hand, if the cost of adapting to the risk is trivial, it might be worth doing even if your chances of benefiting from your action are small. We should all use seat belts, even though few of us will ever be involved in an auto wreck.
- Who decides what is an acceptable level of risk?
- Who is personally at risk? People who live next to the dump? People exposed to radon, but only if they smoke? Everyone?
- What kinds of exposure are important? Note that once someone defines a given exposure level as safe, people tend to assume anything more is dangerous, which may not be the case.
- Can anything be done to reduce risk? What are the risks of the alternative approach? What trade-offs are involved?

How to Evaluate Samples, Surveys, and Polls

Until the 2016 election, the greatest polling fiasco in history was the *Literary Digest* poll in the 1936 presidential election. It predicted the Republican challenger, Alf Landon, would handily defeat the Democratic incumbent, Franklin Delano Roosevelt. In fact, Roosevelt triumphed in a landslide.

How did this happen? The problem was in the way the magazine chose its sample. It relied largely on telephone listings and the roster of its own subscribers. But in 1936, the nation was still in the grip of the Depression. Lots of families did not have extra money to spend on literary magazines or even telephones. The people who did—the people with more money—tended to vote Republican. Result: a tremendous, self-induced black eye for *Literary Digest*.

So, when you are looking at a survey, consider how the sample was chosen—did it come from the population in question, was it chosen by chance? Did every member of the population in question have the same chance of being chosen?

Achieving this standard can be complex. For example, pollsters often conduct telephone polls using dialing programs that dial telephones randomly. Is that random enough? No. The people who answer the phone at the hours the poll is conducted may be people more likely to be at home during those hours. For example, they may be more likely to be older people, or parents. So pollsters may add another question, such as which adult in the household most recently had a birthday, and arrange to talk with that person.

Approach polls with caution if the subjects are, in the language of polling, "self-selected"—if the people volunteer to answer the questions, a major issue with polls conducted online.

The response rate is the number of eligible participants who complete a survey. Researchers have long assumed that higher response rates mean more accurate survey results but, according to the American Association for Public Opinion Research, response rates in recent years have fallen "in some cases precipitously." As a result, polling companies have put more effort into obtaining reliable results, even when participation is not what they would have once regarded as optimal. According to the association, "there is currently no consensus about the factors that produce the disjuncture between response rates and survey quality."[20]

Coming as it does from an organization devoted to polling, this information might cause you to raise a skeptical eyebrow about the entire enterprise. So it is reassuring to see the association offer its own skeptical advice—that consumers of polls pay attention to other indications of polling quality, and note how well the findings fit in with previous research on the issue.

When a survey examines only a sample of a population, there is always a chance that the sample is not really representative of the group, which is to say there is always a chance the poll results will be inaccurate. In general, the larger the sample, the smaller the margin of error—up to a point. According to the association, the margin of sampling error (or MOSE) falls as the number of people polled rises from 100 to 1,000. For example, at 100, the MOSE is 10—so even seemingly large differences might be, statistically speaking, meaningless. By 1,000, though, the MOSE is only 3 percentage points. With a sample of 2,500 it is typically 2 percentage points, and at 5,000 it falls to 1 percentage point. For most purposes, a margin of plus or minus 3 percentage points is adequate. That is why many polls involve a sample of at least 1,000.

Statisticians use a technique called weighting to further refine survey results, adjusting the data so that they more accurately reflect the population from which they were drawn. For example, national telephone surveys typically reach many older women and relatively few younger men. So results from those groups may be artificially altered, or weighted, so that the results may better reflect the entire population.

A self-selected opinion poll, or SLOP survey, is the kind of polling that takes place when someone stands in a shopping mall with a clipboard and a questionnaire, soliciting the opinions of passersby, or when someone posts a questionnaire online and invites people to respond. It is impossible to calculate their margins of sampling error. In fact it is impossible to rely on this kind of polling at all.

Often, people reporting survey results will tell you what the questions were. This information can be illuminating. For example, were participants asked open-ended questions such as, "What should be the aims of the nation's research agenda?" or were they asked to rank the importance of a given list of research activities?

Was the question a leading question—that is, did it associate one alternative or another with an authority figure or widely endorsed goal? Such "push" polls are not polls at all. They are actually a form of telemarketing, and nasty telemarketing at that. You are unlikely to encounter them in a science or engineering context, but you should know what they are just the same.

In the typical push poll you may be asked a few questions about a candidate, but the purpose of the questions ("Are you aware that candidate X did some despicable thing?") is not to ask your opinion or test your knowledge but rather to create a negative impression in your mind.

The American Association for Public Opinion Research regards push-polling as unethical and offers some guidance on figuring out whether you have been called by a push-poller.[21] If the poll has one or more of these features, it may be a push poll:

- You are asked only a few questions, and about a single candidate or issue.
- The questions are uniformly negative. (Sometimes, someone seeking to advance a candidate or issue may do a reverse push poll, in which case the questions will be few, and uniformly positive.)
- If you ask about the poll, your questions draw evasive answers. The poll is not named, or the name is phony (hard to determine on the fly, I realize).

Sometimes, however, what looks like a push poll may be a form of what pollsters call message testing, a way of testing the effectiveness of campaign messages, ads, or the like. In a legitimate message-testing poll the caller will say who is conducting it (though not necessarily who the client is), will ask multiple questions, and will ask you for demographic information—your age, education level, political affiliation, and so on.

So, if you are called by someone purporting to take a poll, ask the name and location of the organization conducting it, how many people will be called, and how the information will be used. If you are not happy with the answers you receive, bow out.

As political campaigns grow increasingly polarizing, even strange and alarming, it has become clear that polling remains important, even in the age of the internet and the cell phone. Among other things, television networks decide who will appear in televised debates according to poll figures. Nevertheless, the polling profession is "teetering on the edge of disaster," as the Harvard historian Jill Lepore said in an address to the Shorenstein Center on Media, Politics, and Public Policy at the Kennedy School of Government.[22] Almost half of Americans no longer have landlines, and restrictions on robo-calls to cell phones, though often violated, have theoretically put them more or less off-limits to responsible pollsters. So polling organizations are turning to online polling, something in the past they rejected because respondents are self-selected and therefore more likely to skew findings. Meanwhile, people are less inclined in general to respond to surveys.

By the end of the primary campaign of 2016, political pros were advising citizens to look not just at one or two polls but the polling history of a race for clues as to what might be going on. News organizations that follow that practice seem to end up with more reliable results.

On the morning of Election Day 2016, analysts were reporting that the Democratic candidate, Hillary Rodham Clinton, had an 80 percent chance of winning the White House. And she did win the popular vote. In states she unexpectedly failed to carry, results were within the margin of sampling error. Overall, though, the polling failure left many people scratching their heads. One popular explanation was that Trump voters, ashamed of supporting a widely reviled candidate, lied to the pollsters.

Be that as it may, remember that, in a way, polls are models of reality, with models' usefulness and deficiencies.

INTRODUCTION

1. C. P. Snow, *The Two Cultures* (Cambridge University Press, 1959, reprinted 1998), p. 60.

2. Ibid., p. 25.

3. Robert Park, *Voodoo Science* (Oxford University Press, 2001), pp. viii–ix.

1. WE THE PEOPLE

1. Claudia Dreifus, *Scientific Conversations* (Holt Paperbacks, 2002).

2. Claudia Dreifus, lecture, New York, January 6, 2003.

3. National Science Board, *Science and Engineering Indicators 2016*, chap. 7, "Science and Technology: Public Attitudes and Understanding"; http://www.nsf.gov/statistics/2016/nsb20161/#/report/chapter-7/; see Appendix Table 7.8.

4. National Commission on Excellence in Education, "A Nation at Risk: The Imperative for Educational Reform," April 1983; http://www2.ed.gov/pubs/NatAtRisk/index.html.

5. "AAAS Science Assessment," Project 2061; http://assessment.aaas.org; "AAAS Testing Web Site Probes Students' Misconceptions about Science," *Science,* April 29, 2011, p. 552.

6. Sam Dillon, "Few Students Show Proficiency in Science, Tests Show," *New York Times,* January 26, 2011.

7. "America's Lab Report: Investigations in High School Science," National Research Council, 2005.

8. Yudhijit Bhattacharjee, "New Curricula Aim to Make High School Labs Less Boring," *Science* 310 (October 14, 2005): 224.

9. Cornelia Dean, "Evolution Takes a Back Seat in U.S. Classes," *New York Times,* February 1, 2005; http://www.nytimes.com/2005/02/01/science/01evo.html.

10. Kevin M. Theissen, "What Do U.S. Students Know about Climate Change?," *EOS* 92, no. 51 (December 20, 2011): 477.

11. Eric Plutzer et al., "Climate Confusion among U.S. Teachers," *Science* (February 12, 2016), p. 664.

12. Kiju Jung et al., "Female Hurricanes Are Deadlier than Male Hurricanes," *Proceedings of the National Academy of Sciences,* June 2, 2014; http://www.pnas.org/content/111/24/8782.short.

13. Robert Park, *Voodoo Science* (Oxford University Press, 2000), p. 35.

14. Amos Tversky and Daniel Kahneman, "Judgment under Uncertainty: Heuristics and Biases," *Science,* September 27, 1974; http://people.hss.caltech.edu/~camerer/Ec101/JudgementUncertainty.pdf.

15. Tversky would certainly have shared it had he not died in 1996; the prize is not awarded posthumously.

16. Dean Buonomano, *Brain Bugs* (W.W. Norton, 2012).

17. There's a similar version, *cum hoc ergo propter hoc,* meaning "with that, therefore because of that."

18. "Fear and Its Consequences," *Scientific American,* February 2011, p. 14.

19. "Freud, Finance and Folly," *Economist,* January 24, 2004, p. 5.

20. See Daniel Simons, "Selective Attention Test"; https://www.youtube.com/watch?v=vJG698U2Mvo.

21. Statement of National Commission on the BP Deepwater Horizon Oil Spill and Offshore Drilling, January 11, 2011.

22. Roger G. Kennedy, *Wildfire and Americans* (Hill & Wang, 2007).

23. February 16, 2011; organized by the Center for Risk Science Innovation and Application.

24. National Academy of Sciences, Woods Hole, MA, August 2–3, 2010.

25. John Allen Paulos, *Innumeracy: Mathematic Illiteracy and Its Consequences* (Farrar, Straus & Giroux, 1989).

26. Donald E. Shelton, "The 'CSI Effect': Does It Really Exist?," *NIJ Journal* 259 (March 2008).

27. Jonathan G. Koomey, *Turning Numbers into Knowledge* (Analytics Press, 2003), pp. 22–23.

28. Marcia E. Herman-Giddens et al., "Secondary Sexual Characteristics and Menses in Young Girls Seen in Office Practice: A Study from the Pediatric Research in Office Settings Network," *Pediatrics* 99, no. 4 (April 1, 1997): 505–512.

29. American College of Obstetricians and Gynecologists, "Menstruation in Girls and Adolescents: Using the Menstrual Cycle as a Vital Sign," *Committee Opinion* 651 (December 2015); http://www.acog.org/Resources-And-Publications/Committee-Opinions/Committee-on-Adolescent-Health-Care/Menstruation-in-Girls-and-Adolescents-Using-the-Menstrual-Cycle-as-a-Vital-Sign.

30. Committee on Adolescence, American Academy of Pediatrics, "Menstruation in Girls and Adolescents: Using the Menstrual Cycle as a Vital Sign," *Pediatrics* 118, no. 5 (November 1, 2006): 2245–2250.

31. Gina Kolata, "In Update on Sperm, Data Show No Decline," *New York Times,* June 7, 2011.

32. U.S. Census Bureau, "Decennial Censuses, 1890 to 1940, and Current Population Survey, Annual Social and Economic Supplements, 1947 to 2015," figure MS-2; http://www.census.gov/hhes/families/files/graphics/MS-2.pdf.

33. National Cancer Institute, Surveillance, Epidemiology, and End Results Program, "SEER Stat Fact Sheets: Brain and Other Nervous System Cancer"; http://seer.cancer.gov/statfacts/html/brain.html.

34. "Fraud Charges Cast Doubt on Claims of DNA Damage from Cell Phone Fields," *Science* 321 (August 29, 2008): 1144.

35. Denis Aydin et al., "Mobile Phone Use and Brain Tumors in Children and Adolescents: A Multicenter Case-Control Study," *Journal of the National Cancer Institute,* August 17, 2011, p. 11.

36. John T. Boice Jr. and Robert E. Tarone, "Cell Phones, Cancer, and Children," *Journal of the National Cancer Institute,* July 27, 2011.

37. Siddhartha Mukherjee, "Patrolling Cancer's Borderlands," *New York Times,* July 17, 2011, SR8.

38. Tara Parker Pope, "New Study Finds that Cellphone Use Changes Some Activity in the Brain," *New York Times,* February 23, 2011, A13.

39. Kate Murphy, "Cellphone Radiation May Alter Your Brain. Let's Talk," *New York Times,* March 31, 2011, B9.

40. Tara Parker-Pope, "Study Finds No Cancer Risk in Cellphones," *New York Times,* October 21, 2011, A6.

41. Nicole Hollander, *Boston Globe,* August 28, 2008, p. 28.

42. Seth Borenstein, "Cellphone Radiation Study Raises Concerns Despite Low Risk," May 27, 2016.

43. Carol Pogash, "Berkeley Offers Safety Guidance on Carrying Phones," *New York Times,* July 24, 2015, A13.

44. Joel M. Moskowitz, "Cellphone Industry Product Liability Lawsuit," *Electromagnetic Radiation Safety* (blog), November 24, 2015; http://www.saferemr.com/2014/08/major-breakthrough-in-cellphone.html.

45. Jon Biao, "Prevalence of Autism Spectrum Disorder among Children Aged 8 Years—Autism and Developmental Disabilities Monitoring Network, 11 Sites, United States, 2010," *Morbidity and Mortality Weekly Report,* Centers for Disease Control and Prevention, March 28, 2014; http://www.cdc.gov/mmwr/preview/mmwrhtml/ss6302a1.htm?s_cid=ss6302a1_w.

46. A. J. Wakefield et al. [retracted], "Ileal-Lymphoid-Nodular Hyperplasia, Nonspecific Colitis, and Pervasive Developmental Disorder in Children," *Lancet* 351 (February 28, 1998): 637–641; http://www.thelancet.com/journals/lancet/article/PIIS0140-6736%2897%2911096-0/abstract.

47. California Department of Public Health, "Pertussis (Whooping Cough)"; http://www.cdph.ca.gov/healthinfo/discond/pages/pertussis.aspx.

48. Biao, "Prevalence of Autism."

49. Pam Belluck and Melena Ryzik, "Tribeca Film Festival to Screen an Anti-Vaccination Movie," *New York Times,* March 21, 2016, A11.

50. Joshua Keating, "If It Happened There: Traditional Beliefs and Distrust of Authority Fueling Disease Outbreak," *Slate,* February 2, 2015.

51. Nicholas A. Christakis, "This Allergies Hysteria Is Just Nuts," *British Medical Journal* 337, no. 7683 (December 13, 2008): 1384.

52. Meredith Broussard, "Everyone's Gone Nuts," *Harpers,* January 2008, p. 64.

53. Letters, *Harpers,* March 2008, p. 6.

54. George Du Toit et al., "Randomized Trial of Peanut Consumption in Infants at Risk for Peanut Allergy," *New England Journal of Medicine* 372 (February 26, 2015): 803–813; http://www.nejm.org/doi/full/10.1056/NEJMoa1414850.

55. Andrew Pollack, "Peanuts as Ally against a Rise in Nut Allergy," *New York Times,* February 24, 2015, A1.

56. "Not So Rash—Reassessing Food Allergy Diagnoses and Treatments," *RAND Review* (Summer 2010): 5 (a study reported in the May 12, 2010 issue of the *Journal of the American Medical Association*).

57. David Ropeik, "The Risk of Poor Coverage of Risk," *Columbia Journalism Review,* November 30, 2010; http://www.cjr.org/the_observatory/the_risk_of_poor_coverage_of_r.php.

58. "Pregnant Women Urged to Eat Fish," *Providence Journal,* June 11, 2014, A9 (*Journal* wire services); Anahad O'Connor, "Health Officials Call for More Fish in Diets of Children and Pregnant Women," *New York Times,* June 11, 2014, A14.

59. Denise Grady, "Cancer Society Criticizes Federal Panel as Overstating Risks," *New York Times,* May 7, 2010, A16.

60. Cristian Tomasetti and Bert Vogelstein, "Variation in Cancer Risk among Tissues Can Be Explained by the Number of Stem Cell Divisions," *Science,* January 2, 2015, pp. 78–81.

61. Kate Kelland, "Special Report: How the World Health Organization's Cancer Agency Confuses Consumers," Reuters, April 18, 2016; http://www.reuters.com/article/us-health-who-iarc-special-report-idUSKCN0XF0RF.

62. T. G. Deryabina et al., "Long-Term Census Data Reveal Abundant Wildlife Populations at Chernobyl," *Current Biology* 25, no. 19 (October 5, 2015): R824–R826; doi: 10.1016/j.cub.2015.08.017.

63. Baruch Fischhoff, "Risk Perception and Health Behavior," in *Cambridge Handbook of Psychology, Health and Medicine,* 2d ed., ed. S. Ayers et al (Cambridge University Press, 2007), pp. 187–191.

64. Andrew C. Revkin, "Behind Toxic Headlines," DotEarth, *New York Times,* January 14, 2011.

65. Ibid.

66. "Guidelines for Investigating Clusters of Health Events," *Morbidity and Mortality Weekly Report,* Centers for Disease Control and Prevention, July 27, 1990, p. 2; https://www.cdc.gov/mmwr/preview/mmwrhtml/00001797.htm.

67. Ibid., p. 8.

68. Ibid., p. 4.

69. Ibid., p. 2.

70. "Bisphenol A (BPA)," Fact Sheet, National Toxicology Program, August 2010; http://www.niehs.nih.gov/research/supported/assets/docs/a_c/bpa_fact_sheet_508.pdf.

71. "Cell Phone Radiofrequency Radiation Studies," Fact Sheet, National Toxicology Program, November 2014; https://www.niehs.nih.gov/health/assets/docs_a_e/cell_phone_radiofrequency_radiation_studies_508.pdf.

2. The Research Enterprise

1. Kenneth R. Miller, personal communication.
2. Francisco Ayala, *Darwin's Gift to Science and Religion* (Joseph Henry Press, 2007).
3. Ibid., p. 11.
4. Understanding Science, "A Blueprint for Scientific Investigations"; http://undsci .berkeley.edu/article/0_0_0/howscienceworks_03.
5. Stephen Jay Gould, *Rocks of Ages* (Ballantine, 1999).
6. Joe Murray, personal communication.
7. Thomas Kuhn, *The Structure of Scientific Revolutions* (University of Chicago Press, 1962), pp. 16–17.
8. Faculty of 1000; http://f1000research.com; F1000 Research / Channels, Preclinical Reproducibility and Robustness; http://f1000research.com/channels/PRR.
9. Jocelyn Kaiser, "Calling All Failed Replication Experiments," *Science*, February 6, 2016, p. 548.
10. "*Pneumocystis* Pneumonia—Los Angeles," *Morbidity and Mortality Weekly Report,* Centers for Disease Control and Prevention, June 5, 1981; http://www.cdc .gov/mmwr/preview/mmwrhtml/june_5.htm.
11. John Allen Paulos, *Innumeracy* (Holt McDougal, 2001), pp. 161–162.
12. John Ioannidis, "Why Most Published Research Findings Are False," *PLoS Medicine,* August 30, 2005; http://dx.doi.org/10.1371/journal.pmed.0020124.
13. Robert Lee Holts, "Most Science Studies Appear to Be Tainted by Sloppy Analysis," *Wall Street Journal*, September 14, 2007, p. B1.
14. "Sloppy Stats Shame Science," *Economist*, June 5, 2004, p. 74.
15. Ramal Moonesinghe et al., "Most Published Research Findings Are False, but a Little Replication Goes a Long Way," *PLoS Medicine,* February 27, 2007; http://dx.doi.org/10.1371/journal.pmed.0040028; also see Benjamin Djulbegovic and Iztok Hozo, "When Should Potentially False Research Findings Be Considered Acceptable?," *PLoS Medicine,* February 27, 2007; http://dx.doi.org/10 .1371/journal.pmed.0040026.
16. Nancy L. Jones, "Raising Scientific Experts," *American Scientist* 99 (November–December 2011): 458.
17. Jonathan Knight, "Null and Void," *Nature* 422 (April 10, 2003): 554.
18. Janet Raloff, "Reviewers Prefer Positive Findings," *Science News*, October 10, 2009, p. 9.
19. Knight, "Null and Void."
20. Sharon Begley, "New Journals Bet 'Negative Results' Save Time, Money," *Wall Street Journal*, September 15, 2006, p. B1.
21. Jonathan Schooler, "Unpublished Results Hide the Decline Effect," *Nature* 470 (February 24, 2011): 437.
22. Orrin H. Pilkey Jr., personal communication, December 23, 2003.
23. Stephen Schneider, "Touch Decisions: Dealing with Uncertainty in Managing Marine Fisheries," presentation at the annual meeting of the American Association for the Advancement of Science, Seattle, February 15, 2004.

24. Orrin H. Pilkey Jr. and Linda Pilkey-Jarvis, *Useless Arithmetic* (Columbia University Press, 2009).

25. The World Meteorological Organization has an excellent explanation at its website; http://www.wmo.int/pages/themes/climate/climate_models.php.

26. "Sea-Ice Models Don't Measure Up," *Nature* 471 (March 2, 2011): 9.

27. Cornelia Dean, "Even Before Its Release, World Climate Report Is Criticized as Too Optimistic," *New York Times,* February 2, 2007; http://www.nytimes.com /2007/02/02/science/02oceans.html.

28. Saul Hansell, "Wall Street Lied to Its Computers," *New York Times*, September 22, 2008, C7.

29. James Hack, quoted in Zeela Merali, "Error: Why Scientific Programming Does Not Compute," *Nature* 467 (October 14, 2010): 773.

30. Nick Barnes, "Publish Your Computer Code: It Is Good Enough," *Nature* 467 (October 14, 2010): 754.

31. Schneider, "Touch Decisions."

32. Euan Nisbet, "Cinderella Science," *Nature* 450 (December 6, 2007): 789.

33. Ibid. Of course, Keeling's work contributed to the work of the IPCC, which won the Nobel Peace Prize in 2007.

34. Cornelia Dean, "Thoreau Is Rediscovered as a Climatologist," *New York Times,* October 28, 2008, D1.

35. Noha Gaber et al., "Guidance on the Development, Evaluation, and Application of Environmental Models," Office of Science Advisor, Council for Regulatory Environmental Modeling, U.S. Environmental Protection Agency, Washington, D.C., March 2009.

36. Schneider, "Touch Decisions."

37. Jonathan G. Koomey, *Turning Numbers into Knowledge: Mastering the Art of Problem Solving,* 2nd ed. (Analytics Press, 2008), p. 128.

38. "Let the Light Shine In," *Economist,* June 14, 2014, pp. 72–73.

39. William F. Perrin, "In Search of Peer Reviewers," *Science,* January 4, 2008, p. 23.

40. Monica Bradford, personal communication, December 15, 2005.

41. Gina Kolata, "Assigning Blame If Fraud Is Found," *New York Times*, September 29, 2002, Week in Review, p. 4.

42. Davis Grimm, "Suggesting or Excluding Reviewers Can Help Get Your Paper Published," *Science*, September 23, 2005, p. 1974.

43. *Answers Research Journal,* Call for Papers; http://www.answersingenesis.org/ari /call-for-papers (expired); now at https://answersingenesis.org/answers/research -journal/about/.

44. Ginger Pinholster, "Journals and Funders Confront Implicit Bias in Peer Review," *Science,* May 27, 2016, pp. 1067–1068.

45. Ibid., p. 1068.

46. Paul Vooshen, "Journal Publishers Rethink a Research Mainstay: Peer Review," *Chronicle of Higher Education* (October 16, 2015): PA10.

47. "Let the Light Shine In."

48. European Space Agency, "Planck: Gravitational Waves Remain Elusive"; http://www.esa.int/Our_Activities/Space_Science/Planck/Planck_gravi tational_waves_remain_elusive.

49. "Einstein Eschews Peer Review," *Science,* March 6, 2015, p. 1032.

50. Dennis Overbye, "Gravitational Waves Detected, Confirming Einstein's Theory," *New York Times,* February 11, 2016; http://www.nytimes.com/2016/02 /12/science/ligo-gravitational-waves-black-holes-einstein.html.

3. THINGS GO WRONG

1. Kenneth Chang, "On Scientific Fakery and the Systems to Catch It," *New York Times,* October 15, 2002, F1.

2. Robert Merton, "The Normative Structure of Science" (1942), collected in Robert Merton, *The Sociology of Science* (University of Chicago Press, 1973), p. 268.

3. Ibid., pp. 274–275.

4. Sandra Titus and Xavier Bosch, "Tie Funding to Research Integrity," *Nature* 466 (July 22, 2010): 436. Bosch is at the University of Barcelona, Spain; Titus, at the Office of Research Integrity of the U.S. Department of Health and Human Services.

5. Gerald Koocher and Patricia Keith-Spiegel, "Peers Nip Misconduct in the Bud," *Nature* 466 (July 22, 2010): 438–440.

6. "Hauser Report Released," *Science,* June 6, 2014, p. 1068.

7. Brian C. Martinson et al., "Scientists Behaving Badly," *Nature* 435 (June 9, 2005): 737.

8. Harvard School of Public Health, "Survey of Academic Medical Centers' Agreements with Industry Finds Differing Standards for Control of Clinical Trial Results," May 25, 2005; http://archive.sph.harvard.edu/press-releases/archives /2005-releases/press05252005.html. Cornelia Dean, "Medical Schools Found to Vary in Their Drug-Testing Standards," *New York Times*, May 26, 2005, p. A24.

9. "Under Suspicion," *Nature* 464 (April 29, 2010): 1245.

10. David Tuller, "Fatigue Syndrome Study Is Retracted by Journal," *New York Times*, December 23, 2011, p. A17.

11. "Second Paper Pulled on Viral Link to CFS," *Science,* January 6, 2012, p. 18.

12. Adam Marcus and Ivan Oransky, "The Paper Is Not Sacred," *Nature* 480 (December 22–29, 2011): 449. The Retraction Watch website: retractionwatch.com.

13. Michael L. Grienelsen and Minghua Zhang, "A Comprehensive Survey of Retracted Articles from the Scholarly Literature," *PLoS One*, October 2012; http:// journals.plos.org/plosone/article?id=10.1371/journal.pone.0044118.

14. Ferric C. Fang et al., "Misconduct Accounts for the Majority of Retracted Scientific Publications," *PNAS,* October 2012.

15. C. Glenn Begley and Lee M. Ellis, "Drug Development: Raise Standards for Preclinical Cancer Research," *Nature* 483 (March 2012): 531–533; Monya Baker, "First Results From Psychology's Largest Reproducibility Test," *Nature*, April 30,

2015; doi:10.1038/nature.2015.17433, reported in Benedict Carey, "Psychology's Fears Confirmed: Rechecked Studies Don't Hold Up," *New York Times,* August 28, 2015, p. A1. Almost immediately, however, subsequent research suggested this finding might be a bit too pessimistic!

16. Andrew C. Chang and Phillip Li, "Is Economics Research Replicable? Sixty Published Papers from Thirteen Journals Say "Usually Not," Finance and Economics Discussion Series 2015-083, Washington: Board of Governors of the Federal Reserve System; http://dx.doi.org/10.17016/FEDS.2015.083.

17. Richard Van Noorden, "The Trouble with Retractions," *Nature* 478 (October 6, 2011): 27.

18. CrossMark; www.crossref.org/crossmark.

19. "Do Retractions Work?," *Nature* 478 (October 6, 2011): 28.

20. Marcia McNutt, "Due Process in the Twitter Age," *Science,* April 22, 2016, p. 387.

21. John Schwartz, "Family's Effort to Clear Name Frames Debate on Executions," *New York Times*, October 15, 2010, p. A17.

22. *Strengthening Forensic Science in the United States* (National Academy Press, 2009).

23. C. W. Griffin, "Warning: Scientific Evidence Can Be Soporific," *Washington Post,* December 11, 1996.

24. Adam Liptak, "Eyewitness Evidence Needs No Special Cautions, Court Says," *New York Times,* January 12, 2012, p. A3.

25. New Jersey Courts, "Supreme Court Releases Eyewitness Identification Criteria for Criminal Cases"; July 19, 2012; https://www.judiciary.state.nj.us/pressrel/2012 /pr120719a.htm.

26. David M. Studdert, Michelle M. Mello, and Troyen A. Brennan, "Medical Malpractice," *New England Journal of Medicine* 350 (January 15, 2004): 283–292; http://www.nejm.org/doi/full/10.1056/NEJMhpr035470.

27. *Frye v. United States,* 54 App. D.C. at 47, 293 F., at 1014.

28. Federal Rules of Evidence #401.

29. Ibid., #702.

30. 509 U.S. 579, pp. 92–102.

31. Ibid., pp. 5–6.

32. Janet Raloff, "Judging Science," *Science News*, January 19, 2008.

33. "Evaluation of the Reference Manual on Scientific Evidence," National Research Council (National Academies Press, 2009).

34. Sheila Jasanoff, *Science at the Bar* (Harvard University Press, 1997).

35. Cornelia Dean, "When Questions of Science Come to a Courtroom, Truth Has Many Faces," *New York Times*, December 5, 2006, p. F3.

36. Presentation at the Shorenstein Center on the Press, Politics and Public Policy, Kennedy School of Government, Harvard University, October, 21, 2003.

37. Jeff Burnside, personal communication, Pew Conference, October 15, 2003.

38. James Glanz, "Cut the Communications Fog, Say Physicists and Editors," *Science,* August 15, 1997, p. 895.

39. "Blondes 'to Die Out in 200 Years,'" *BBC News,* Sept. 27, 2002; http://news.bbc.co.uk//1/hi/health/2284783.stm.

40. "Extinction of Blondes Vastly Overreported," *Washington Post,* October 2, 2002, p. C1.

41. Cornelia Dean, "Rousing Science Out of the Lab and Into the Limelight," *New York Times,* November 11, 2003; http://www.nytimes.com/2003/11/11/science/commentary-rousing-science-out-of-the-lab-and-into-the-limelight.html.

42. Dylan Byers, "Aaron Sorkin: Media Biased toward Fairness," *Politico,* June 16, 2012; http://www.politico.com/blogs/media/2012/06/aaron-sorkin-media-biased-toward-fairness-126357.

43. "Ersatz Eve," *New York Times,* December 28, 2002, p. A18.

44. Todd S. Purdum, "Washington, We Have a Problem," *Vanity Fair,* September 2010, p. 337.

45. Joel Achenbach, "Reality Check," *Washington Post,* December 4, 1996, p. C1.

46. Jim Giles, "Internet Encyclopaedias Go Head to Head," *Nature* 438 (December 15, 2005); http://www.nature.com/nature/journal/v438/n7070/full/438900a.html.

47. Adam Gopnik, "The Information," *New Yorker,* February 14 and 21, 2011, p. 126.

48. Alan Alda, personal communication, New York City, May 27, 2003.

4. THE UNIVERSAL SOLVENT

1. Richard Carter, *Breakthrough: The Saga of Jonas Salk* (Trident Press, 1966), p. 283.

2. Robert Merton, "The Normative Structure of Science" (1942), collected in Robert Merton, *The Sociology of Science* (University of Chicago Press, 1973).

3. "The wise man has a country; science does not."

4. Merton, "The Normative Structure of Science," pp. 270–271.

5. Ben Ritz, "Renewing Federal Investment in Research and Development," May 25, 2016; http://bipartisanpolicy.org/blog/renewing-federal-investments-in-research-and-development/.

6. "The Bayh-Dole Act: A Guide to the Law and Implementing Regulations," Council on Governmental Relations, October 1999.

7. Daniel Sarewitz, "Science Agencies Must Bite Innovation Bullet," *Nature* 4712 (March 10, 2011): 137.

8. Council on Governmental Relations, "The Bayh-Dole Act," p. 10.

9. Henry Thomas Stelfox et al., "Conflict of Interest in the Debate over Calcium-Channel Antagonists," *New England Journal of Medicine* 383, no. 2 (January 8, 1998): 101–106.

10. Justin E. Bekelman et al., "Scope and Impact of Financial Conflicts of Interest in Biomedical Research," *Journal of the American Medical Association* 289, no. 4 (January 22/29, 2003): 454–465.

11. Merton, "The Normative Structure of Science," p. 273.

12. Sheldon Krimsky, *Science in the Private Interest* (Rowman & Littlefield, 2003), p. 111.

13. Erik Stockstad, "University Bids to Salvage Reputation after Flap over Logging Paper," *Science*, June 2, 2006, p. 1288.

14. D. C. Donato et al., "Post-Wildfire Logging Hinders Regeneration and Increases Fire Risk," *Science*, January 20, 2006, p. 352; http://www.sciencemag.org/content /311/5759/352; Stockstad, "University Bids to Salvage Reputation."

15. Union of Concerned Scientists, "BLM Cuts Forest Grant after Study Criticizes Tree Cutting," http://www.ucsusa.org/center-for-science-and-democracy/scientific _integrity/abuses_of_science/a-to-z/post-disturbance-logging.html.

16. Krimsky, *Science in the Private Interest,* p. 224.

17. Don A. Moore et al., "Conflict of Interest and the Unconscious Intrusion of Bias," Social Science Research Network Electronic Library, Harvard Working Paper No. 02–40.

18. Lenard I. Lesser et al., "Relationship between Funding Source and Conclusion among Nutrition-Related Scientific Articles," *PLoS Medicine* (January 2007): E5.

19. "Panel Pans UC-Novartis Deal," *Science,* July 30, 2004, p. 591.

20. "Berkeley Grants Tenure to Critic," *Chronicle of Higher Education,* June 3, 2005, A8.

21. Andrew Delbanco, "In Memoriam," *New York Review of Books,* February 27, 2003, p. 23; quotation on p. 24.

22. Yu Xie, "Undemocracy: Inequalities in Science," *Science* 344 (May 25, 2014): 809.

23. Mori Matsumoto et al., "MRI of Cervical Intervertebral Discs in Asymptomatic Subjects," *Journal of Bone & Joint Surgery,* January 1, 1998.

24. James N. Weinstein et al., "United States Trends and Variations in Regional Lumbar Spine Surgery: 1992–2003," *Spine* 31 (November 2006): 2707–2714; http://www.ncbi.nlm.nih.gov/pmc/articles/PMC2913862/.

25. Gina Kolata, "Sports Medicine Said to Overuse a Popular Scan," October 29, 2011, p. A1.

26. J. Bruce Moseley, M.D. et al., "A Controlled Trial of Arthroscopic Surgery for Osteoarthritis of the Knee," *New England Journal of Medicine* 347 (July 11, 2002): 81–88.

27. Barnett S Kramer et al., "Lung Cancer Screening with Low-Dose Helical CT: Results from the National Lung Screening Trial (NLST)," *Journal of Medical Screening* 18 (September 2011): 109–111.

28. National Cancer Institute, "National Lung Screening Trial"; http://www.cancer .gov/clinicaltrials/noteworthy-trials/nlst.

29. Lung Cancer Foundation of America, "The National Lung Screening Trial (NLST) Results Regarding Early Detection"; http://www.lcfamerica.org/the -national-lung-screening-trial-nlst-results-regarding-early-detection/.

30. American Cancer Society, "Breast Cancer Facts and Figures 2013–2014"; http://www.cancer.org/acs/groups/content/@research/documents/document /acspc-042725.pdf.

31. E. Olson and Peter C. Gøtzsche, "Cochrane Review on Screening for Breast Cancer with Mammography," *Lancet* 358 (October 20, 2001): 1340–1342.

32. Peter Gøtzsche, "Mammography Screening Ten Years On: Reflections on a Decade since the 2001 Review," Archive of the Cochrane Community Site; http://www.cochrane.org/news/blog/mammography-screening-ten-years -reflections-decade-2001-review.

33. Mary Roach, *Gulp* (W.W. Norton, 2013), p. 16.

34. Scott O. Lilienfeld, James M. Wood, and Howard N. Garb, "The Scientific Status of Projective Techniques," *Psychological Science in the Public Interest* 1 (November 2000): http://www.therapiebreve.be/documents/lilienfeld-ea-2000 .pdf.

35. Erica Goode, "What's in an Inkblot? Some Say Not Much," *New York Times,* February 20, 2001.

36. William M. Grove et al., "Failure of Rorschach-Comprehensive-System-Based Testimony to Be Admissible under the *Daubert-Joiner-Kumho* Standard," *Psychology, Public Policy, and Law* 8 (2002): 216–234.

37. Peter Orszag, "Malpractice Methodology," *New York Times,* October 21, 2010, p. A19.

38. Lonnie Hanauer, letter to the editor, September 1, 2015, p. A20.

39. "German Doctors Are Told to Have an Open Attitude to Placebos," *BMJ* (2011); http://www.bmj.com/content/342/bmj.d1535.

40. Associated Press, "German Medical Group Pushes Placebo," published in the *Boston Globe,* April 1, 2011, p. A5.

41. Andrew I. Geller et al., "Emergency Department Visits for Adverse Events Related to Dietary Supplements," *New England Journal of Medicine* 373 (October 2015): 1531–1540; http://www.nejm.org/doi/full/10.1056/NEJMsa1504267.

42. Ibid.; Anahad O'Connor, "Dietary Supplements Cause 20,000 E.R. Visits Each Year, Study Says," *New York Times*, October 15, 2015, p. A21.

43. National Center for Complementary and Integrative Health, "Homeopathy"; http://nccam.nih.gov/health/homeopathy.

44. Ibid.

45. Linda Rosa et al., "A Close Look at Therapeutic Touch," *Journal of the American Medical Association* 279 (April 1, 1998): 1005–1010; http://jama.jamanetwork .com/article.aspx?articleid=187390. The fourth grader, Emily Rosa, is believed to be the youngest person ever to report findings in a peer-reviewed medical journal.

46. Ray Moynihan et al., "Coverage by the News Media of the Benefits and Risks of Medications," *New England Journal of Medicine* 342 (June 1, 2000): 1645–1650.

47. Bulletin of the World Health Organization, "Direct-to-Consumer Advertising Under Fire"; http://www.who.int/bulletin/volumes/87/8/09-040809/en/.

48. Ronni Sandroff, "Drugs as a Last Resort," *Consumer Reports,* March 2005.

49. John La Puma, "Don't Ask Your Doctor about 'Low T,'" *New York Times,* February 4, 2014.

50. U.S. Food and Drug Administration, "Key Points of the Bad Ad Program"; http://www.fda.gov/Drugs/GuidanceComplianceRegulatoryInformation

/Surveillance/DrugMarketingAdvertisingandCommunications/ucm211498
.htm.

51. FiercePharma, "Kim Kardashian West Found Morning Sickness Relief with Diclegis"; http://www.fiercepharmamarketing.com/press-releases/kim-kardashian-west-found-morning-sickness-relief-diclegis.

52. Lawrence K. Altman, M.D., "The Doctor's World: Decline in Autopsies Raises Concern," *New York Times,* September 13, 1983.

53. Darin L.Wolfe, "To See for Oneself," *American Scientist,* May–June 2010, p. 228.

54. Altman, "The Doctor's World."

55. Wolfe, "To See for Oneself," p. 234.

56. Keith Roach, "Unnecessary Tests Can Do More Harm than Good," *Providence Journal,* November 3, 2015, p. C6.

57. Abigail Zuger, "When Good Medicine Mixes with Bad," *New York Times,* January 11, 2016; http://well.blogs.nytimes.com/2016/01/11/when-good-medicine-mixes-with-bad/.

58. "Do You Need that Scan?" *Consumer Reports on Health,* September 2011, p. 4.

59. U.S. Department of Agriculture Economic Research Service, "Food Prices and Spending"; http://www.ers.usda.gov/data-products/ag-and-food-statistics-charting-the-essentials/food-prices-and-spending.aspx.

60. www.oversizecasket.com.

61. Office of Disease Prevention and Health Promotion, "Scientific Report of the 2015 Dietary Guidelines Advisory Committee"; http://www.health.gov/dietaryguidelines/2015-scientific-report/02-executive-summary.asp.

62. "Food Pyramid in Need of Renovation," *Harvard Women's Health Watch,* February 2003.

63. Mark Bittman, "My Dream Food Label," *New York Times,* October 13, 2012, p. SR6; http://www.nytimes.com/2012/10/14/opinion/sunday/bittman-my-dream-food-label.html.

64. Stanford Medicine News Center, "Little Evidence of Health Benefits from Organic Foods, Study Finds," September 3, 2012; https://med.stanford.edu/news/all-news/2012/09/little-evidence-of-health-benefits-from-organic-foods-study-finds.html.

65. U.S. Department of Agriculture Economic Research Service, "Organic Agriculture: Organic Market Overview"; http://www.ers.usda.gov/topics/natural-resources-environment/organic-agriculture/organic-market-overview.aspx.

66. The report, "A Science-Based Look at Genetically Engineered Crops," is available here: https://nas-sites.org/ge-crops/.

67. National Health and Nutrition Examination Survey; http://www.cdc.gov/nchs/data/nhanes/survey_content_99_16.pdf.

68. Melanie Warner, "For Corn Syrup, the Sweet Talk Gets Harder," *New York Times,* May 2, 2010, p. WB8.

69. "An -Ose Is an -Ose," *New York Times,* September 15, 2010, p. A23.

70. Candace Choi, Associated Press, "The Latest Weapon in the War on Obesity?" *Providence Journal*, November 25, 2015.

71. Margaret Sanger-Katz, "Yes, Soda Taxes Do Seem to Discourage Soda Drinking," *New York Times,* October 13, 2015, p. A3.

72. The Sugar Association, https://www.sugar.org/about-us/.

73. "Snake Oil in the Supermarket," *Scientific American,* September 2010, p. 30.

74. www.yaleruddcenter.org/what_we_do.aspx?id=4. The center moved to the University of Connecticut in 2015.

75. Federal Trade Commission, "A Review of Food Marketing to Children and Adolescents—Follow-Up Report," December 2012.

76. According to the *New York Times,* the WHO has assessed more than 900 possible carcinogens, of which only one—a chemical used in the manufacturing of nylon that can find its way into drinking water—was declared innocent. Its rogues' gallery of carcinogenic foods also contains pickled vegetables and coffee.

77. Nicholas St. Fleur, "Though Labeled 'Wild,' that Serving of Salmon May Be Farmed or 'Faux,'" *New York Times,* October 29, 2015, p. A24.

78. The Marine Stewardship Council website at http://www.msc.org/tracks the health of fish stocks around the world and recommends for purchase fish and shellfish "harvested in a sustainable manner." The site also offers recipes.

79. Cynthia Graber, "Michael Pollan Explains What's Wrong with the Paleo Diet," *Mother Jones,* January 17, 2014; http://www.motherjones.com/print/243291.

80. N. R. Kleinfield, "Just What Killed the Diet Doctor, and What Keeps the Issue Alive?" *New York Times,* February 11, 2004; http://www.nytimes.com/2004/02/11/nyregion/just-what-killed-the-diet-doctor-and-what-keeps-the-issue-alive.html.

81. U.S. Department of Agriculture, "Dietary Guidelines for Americans, 2015–2020"; https://www.choosemyplate.gov/dietary-guidelines.

82. Mary Clare Jalonick, Associated Press, "Meat Producers Propose Dietary Guidelines," *Providence Journal,* March 23, 2015, p. C4.

83. American Meat Institute, "Meat MythCrushers"; http://meatmythcrushers.com/about.php.

84. Nina Teicholz, "The Government's Bad Diet Advice," *New York Times,* February 20, 2015; http://www.nytimes.com/2015/02/21/opinion/when-the-government-tells-you-what-to-eat.html?_r=0.

5. Political Science

1. Jennifer E. Manning, "Membership of the 114th Congress: A Profile," Congressional Research Service, December 1, 2015; http://www.senate.gov/CRSReports/crs-publish.cfm?pid=%260BL*RLC2%0A.

2. The story of the Montreal Protocol is often held up as an example of good science driving good policy. Many people wonder why the international will that brought it about cannot be brought to bear on issues of climate change. In my opinion, the answer is simple: companies that were going to have to stop selling banned aerosol and other products already had in their marketing pipelines an array of substitutes. As a result, the commercial disruption was minimal.

3. Sherwood Boehlert, personal communication, 2006.

4. Arthur L. Kellermann et al., "Gun Ownership as a Risk Factor for Homicide in the Home," *New England Journal of Medicine* 329 (October 7, 1993): 1084–1091.

5. Arthur L. Kellermann and Frederick P. Rivara, "Silencing the Science on Gun Research," *Journal of the American Medical Association* 309 (February 13, 2013): 549–550.

6. Michael Luo, "Sway of N.R.A. Blocks Studies, Scientists Say," *New York Times*, January 26, 2011, p. A1.

7. Meghan Rosen, "Misfires in the Gun Control Debate," *Science News*, May 14, 2016, pp. 16ff.

8. Cassandra K. Crifasi et al., "Effects of Changes in Permit-to-Purchase Handgun Laws in Connecticut and Missouri on Suicide Rates," *Preventive Medicine* 79 (October 2015): 43–49; http://www.sciencedirect.com/science/article/pii /S0091743515002297.

9. Robert L. Park, *Voodoo Science: The Road from Foolishness to Fraud* (Oxford University Press, 2000); quotations on pp. 186, 183, 188.

10. Daniel S. Greenberg, *Science, Money and Politics* (University of Chicago Press, 2001), p. 280.

11. Cornelia Dean, "Groups Call for Scientists to Engage the Body Politic," *New York Times,* August 8, 2011; http://www.nytimes.com/2011/08/09/science/09emily .html?_r=0.

12. Gus Speth, personal communication.

13. Richard Somerville and Susan Joy Hassol, "Communicating the Science of Climate Change," *Physics Today,* October 2011, pp. 48–53.

14. John Travis, "Inside the Summit on Human Gene Editing: A Reporter's Notebook," *Science,* December 4, 2015; http://www.sciencemag.org/news/2015/12 /inside-summit-human-gene-editing-reporter-s-notebook.

15. James K. Hammitt, Harvard Law School conference on risk prevention, March 10, 2004.

16. Presentation to Nieman Fellows, Harvard University, May 14, 2004.

17. Ibid.

18. National Academy of Sciences, Organized Collections, Committees on Biological Effects of Atomic Radiation, 1954–1964; http://www.nasonline.org/about -nas/history/archives/collections/cbear-1954-1964.html.

19. SciDevNet, "Famine-Stricken Countries Reject GM Maize," July 29, 2002; http://www.scidev.net/global/gm/news/faminestricken-countries-reject-gm -maize.html.

20. Robert E. Neustadt and Ernest R. May, *Thinking in Time: The Uses of History for Decision-Makers* (Free Press, 1988).

21. Paul Harremöes et al., eds., *The Precautionary Principle* (Earthscan, 2002), p. 201.

22. Pew Research Center, "America's Changing Religious Landscape," May 12, 2015; http://www.pewforum.org/2015/05/12/americas-changing-religious-landscape/.

23. National Science Foundation, *Science and Engineering Indicators,* chapter 7; http://www.nsf.gov/statistics/seind14/index.cfm/chapter-7.

24. Francisco Ayala, *Darwin's Gift to Science and Religion* (Joseph Henry Press, 2007), p. 54.

25. Kathrin F. Stanger-Hall and David W. Hall, "Abstinence-Only Education and Teen Pregnancy Rates: Why We Need Comprehensive Sex Education in the U.S.," *PLoS,* October 24, 2011; http://journals.plos.org/plosone/article?id=10 .1371/journal.pone.0024658.

26. Lawrence Selden, "Cornelia's Creed Becomes Famous," *Darwinian Fundamentalism* (blog), October 3, 2007; http://darwinianfundamentalism.blogspot.com /2007/10/cornelias-creed-becomes-famous.html.

27. Ayala, *Darwin's Gift,* p. xiii.

28. National Center for Science Education, "The Wedge Document," posted October 14, 2008; http://ncse.com/creationism/general/wedge-document.

29. Discovery Institute, "'The Wedge Document': 'So What?'"; http://www .discovery.org/f/349.

30. Nicholas J. Matzke, "The Evolution of Creationist Movements," *Evolution: Education and Outreach* 3 (2010): 145; doi:10.1007/s12052-010-0233-1.

31. Cornelia Dean, "Evolution Takes a Back Seat in U.S. Classes," *New York Times,* February 1, 2005; http://www.nytimes.com/2005/02/01/science/01evo.html.

32. Michael B. Berkman and Eric Plutzer, "Enablers of Doubt: How Future Teachers Learn to Negotiate the Evolution Wars in Their Classrooms," *Annals of the American Academy of Political and Social Science* 658 (March 2015): 253–270; http://ann.sagepub.com/content/658/1/253.abstract.

33. *Lemon v. Kurtzman,* 403 U.S. 602 (1971); http://www.oyez.org/cases/1970-1979 /1970/1970_89.

34. Laurie Goodstein, "Judge Rejects Teaching Intelligent Design," *New York Times,* December 21, 2005; http://www.nytimes.com/2005/12/21/education /21evolution.html.

35. *County of Allegheny v. American Civil Liberties Union,* Greater Pittsburgh Chapter, 492 U.S. 573 (1989); http://www.oyez.org/cases/1980-1989/1988/1988_87_2050.

36. Cornelia Dean, "Creationism and Science Clash at Grand Canyon Bookstore," *New York Times,* October 26, 2004.

37. Guy Consolmagno and George Coyne, "Asteroids, Stars, and the Love of God," *On Being, with Krista Tippett,* September 24, 2015; http://www.onbeing.org /program/asteroids-stars-and-love-god/68.

Conclusion

1. *Wollschlaeger v. Governor of Florida*; http://media.ca11.uscourts.gov/opinions /pub/files/201214009.pdf. Also see "Physicians Petition 11th Circuit for Rehearing of Florida Gun Law Case," *American Academy of Pediatrics*; https://www .aap.org/en-us/advocacy-and-policy/state-advocacy/Pages/AmicusFL2014 .aspx.

2. Florida Center for Investigative Reporting, "In Florida Officials Ban Term 'Climate Change,'" March 8, 2015; http://fcir.org/2015/03/08/in-florida-officials-ban-term

-climate-change/; http://www.miamiherald.com/news/state/florida/article12983720
.html.

3. Philip A. Janquart, "Groups Call Wyoming Ag-Gag Law Censorship," *Court-house News Service,* October 1, 2015; http://www.courthousenews.com/2015/10
/01/groups-call-wyoming-ag-gag-law-censorship.htm.

4. A. A. Rosenberg et al., "Congress's Attacks on Science-Based Rules," *Science,* May 29, 2015, pp. 964–966.

5. John Markoff, "Autonomous Weapons' Safety Is Questioned," *New York Times,* Feb. 29, 2016, p. B7.

6. Paul Scharre, "Autonomous Weapons and Operational Risk," report of the Center for a New American Security, February 29, 2016; http://www.cnas.org
/autonomous-weapons-and-operational-risk#.V1dMd2TF-aE.

7. "Ocean Fertilization," Woods Hole Oceanographic Institution; http://www
.whoi.edu/ocb-fert/page.do?pid=38315.

8. Cornelia Dean, "Guidelines for Epidemics: Who Gets a Ventilator?," *New York Times,* March 25, 2008; http://www.nytimes.com/2008/03/25/health/25vent
.html.

9. David Baltimore et al., "A Prudent Path Forward for Genomic Engineering and Germline Gene Modification," *Science,* April 3, 2015, p. 36.

10. Jennifer Doudna, "Embryo Editing Needs Scrutiny," *Nature* 528 (December 3, 2015): 56.

11. Andrew Pollack, "Private Talks Are Conducted about a Synthetic Genome," *New York Times,* May 14, 2016; p. A11.

12. *Gene Drives on the Horizon,* Gene-Drive Research in Non-human Organisms: Recommendations for Responsible Conduct, The National Academies of Sciences, Engineering, and Medicine; http://nas-sites.org/gene-drives/.

13. Megan J. Palmer, Francis Fukuyama, and David A. Relman, "A More Systematic Approach to Biological Risk," *Science,* December 18, 2015, p. 1471.

14. Rosenberg et al., "Congress's Attacks on Science-Based Rules," p. 964.

15. Daniel Sarewitz, "Science Can't Solve It," *Nature* 522 (June 25, 2015): 413–414.

16. Ibid.

17. Harvey V. Fineberg, "Wider Attention for GOF Science," *Science,* February 21, 2015, p. 929.

18. Really what it means, in the United States anyway, is thorough campaign financing reform. In its absence, everything else is futile.

19. News of one such study is available here: "What You Know Depends on What You Watch: Current Events Knowledge Across Popular News Sources," Public Mind Poll, Fairleigh Dickinson University; http://publicmind.fdu.edu/2012
/confirmed/final.pdf. It found misinformation among viewers of MSNBC as well.

Appendix. Trustworthy, Untrustworthy, or Irrelevant?

1. Cara Buckley, "Doubts Rise on Bedbug-Sniffing Dogs," *New York Times,* November 10, 2010.

2. Harry Collins and Robert Evans, *Rethinking Expertise* (University of Chicago Press, 2007), p. 2.

3. He made this point on the *Charlie Rose* show on August 13, 2009. The video is no longer on the show's website but can be viewed here: http://edge.org/news/a-conversation-with-theoretical-physicist-and-mathematician-freeman-dyson.

4. Kenneth Brower, "The Danger of Cosmic Genius," *Atlantic,* December 2010, p. 51.

5. Gina Kolata, "Hope in the Lab: A Special Report," *New York Times,* May 3, 1998; http://www.nytimes.com/1998/05/03/us/hope-lab-special-report-cautious-awe-greets-drugs-that-eradicate-tumors-mice.html.

6. Massimo Pigliucci, *Nonsense on Stilts* (University of Chicago Press, 2010), p. 281.

7. Richard Van Noorden, "Metrics: A Profusion of Measures," *Nature* 465 (June 17, 2010): 864.

8. Ibid., 865.

9. Cornelia Dean, "Believing Scripture but Playing by Science's Rules," *New York Times,* February 12, 2007; http://www.nytimes.com/2007/02/12/science/12geologist.html?_r=0.

10. Pigliucci, *Nonsense on Stilts*, p. 90.

11. Ibid., p. 285.

12. Jonathan G. Koomey, "Sorry, Wrong Number," *IEEE Spectrum* (June 2003): 11–12.

13. Gina Kolata, "Assigning Blame If Fraud Is Found," *New York Times,* Week in Review, September 29, 2005, p. 5.

14. Bruce Alberts, "Promoting Scientific Standards," *Science,* January 1, 2010, p. 12.

15. Judith L. Turgeon et al., "Hormone Therapy: Physiological Complexity Belies Therapeutic Simplicity," *Science,* May 28, 2004, p. 1269.

16. "Notes from the Science Lab"; http://www2.copydesk.org/hold/words/science.htm.

17. In March 2016, the American Statistical Association released a statement suggesting that the research community's death grip on the p-value standard may be having unwelcome effects; https://www.amstat.org/newsroom/pressreleases/P-ValueStatement.pdf.

18. Rui Wang et al., "Statistics in Medicine—Reporting Subgroup Analyses in Clinical Trials," *New England Journal of Medicine* 357 (November 22, 2007): 2189.

19. David Kent and Rodney Hayward, "When Averages Hide Individual Differences in Clinical Trials," *American Scientist* 95 (January-February 2007): 60.

20. American Association for Public Opinion Research (AAPOR), "Poll and Survey FAQ"; http://www.aapor.org/Education-Resources/For-Researchers/Poll-Survey-FAQ.aspx.

21. AAPOR, "Statements on 'Push' Polls"; http://www.aapor.org/Standards-Ethics/Resources/AAPOR-Statements-on-Push-Polls.aspx.

22. Jill Lepore, Theodore H. White Lecture on Press and Politics with Jill Lepore, Shorenstein Center, Harvard Kennedy School, November 5, 2015, p. 14; http://shorensteincenter.org/wp-content/uploads/2015/11/Theodore-H-White-Transcript-2015.pdf.

FURTHER READING

Agin, Dan. *Junk Science*. St. Martin's Press, 2006.

Angell, Marcia, M.D. *Science on Trial: The Clash of Medical Evidence and the Law in the Breast Implant Case*. W.W. Norton & Co., 1996.

Ariely, Dan. *Predictably Irrational: The Hidden Forces that Shape Our Decisions*. HarperCollins, 2008.

Atkins, Robert C., M.D. *Dr. Atkins' New Diet Revolution*. Avon Books, 1992.

Augustine, Norman R. *Is America Falling Off the Flat Earth?* National Academies Press, 2007.

Augustine, Norman R., chair, Committee on Prospering in the Global Economy of the 21st Century. *Rising Above the Gathering Storm: Energizing and Employing America for a Brighter Economic Future*. National Academies Press, 2007.

Ayala, Francisco J. *Darwin's Gift to Science and Religion*. Joseph Henry Press, 2007.

Ayala, Francisco J., chair, Committee on Revising Science and Creationism. *Science, Evolution and Creationism*. National Academies Press, 2008.

Baron, Nancy. *Escape From the Ivory Tower*. Island Press, 2010.

Bausell, R. Barker. *Snake Oil Science*. Oxford University Press, 2007.

Berezow, Alex B. and Hank Campbell. *Science Left Behind: Feel-Good Fallacies and the Rise of the Anti-Scientific Left*. Public Affairs, 2012.

Bridger, Sarah. *Scientists at War: The Ethics of Cold War Weapons Research*. Harvard University Press, 2015.

Brownson, Ross C. et al. *Evidence-Based Public Health*. Oxford University Press, 2003.

Centers for Disease Control. "Guidelines for Investigating Clusters of Health Events." *Morbidity and Mortality Weekly Report* 39 (July 27, 1990): 1–16.

Cohn, Victor. *News and Numbers*. Iowa State University Press, 1989.

Collins, Harry and Robert Evans. *Rethinking Expertise*. University of Chicago Press, 2007.

"Countermeasures: A Technical Evaluation of the Operational Effectiveness of the Planned US National Missile Defense System." Union of Concerned Scientists and the MIT Security Studies Program, April 2000.

Faigman, David L. *Laboratory of Justice*. Times Books, 2004.

———. *Legal Alchemy: The Use and Misuse of Science in the Law*. W.H. Freeman & Co., 1999.

Ferguson, Charles D. *Nuclear Energy*. Oxford University Press, 2011.

Fitzgerald, Frances. *Way Out There in the Blue*. Simon & Schuster, 2000.

Forbes, Kenneth R. and Peter Huber. *Judging Science*. MIT Press, 1997.

Frank, Thomas. *What's the Matter with Kansas.* Metropolitan Books, 2004.

Gladwell, Malcolm. *Blink.* Little Brown & Co., 2005.

Goldacre, Ben. *Bad Science: Quacks, Hacks and Big Pharma Flacks.* Faber & Faber, 2010.

Goldstein, Inge F. and Martin Goldstein. *How Much Risk?* Oxford University Press, 2002.

Gould, Stephen J. *Rocks of Ages.* Ballantine Books, 1999.

Greenberg, Daniel S. *Science, Money and Politics.* University of Chicago Press, 2001.

———. *Science for Sale: The Perils, Rewards, and Delusions of Campus Capitalism.* University of Chicago Press, 2007.

Hallinan, Joseph T. *Why We Make Mistakes.* Broadway Books, 2009.

Harremoës, Paul et al., eds. *The Precautionary Principle in the 20th Century.* Earthscan, 2002.

Hartz, Jim and Rick Chappell. "Worlds Apart: How the Distance between Science and Journalism Threatens America's Future." First Amendment Center, Nashville, TN, 1997.

Horgan, John. *The End of Science.* Addison-Wesley, 1996.

Jasanoff, Sheila. *Science at the Bar.* Harvard University Press, 1995.

Kabat, Geoffrey C. *Hyping Health Risks: Environmental Hazards in Daily Life and the Science of Epidemiology.* Columbia University Press, 2008.

Kahneman, Daniel. *Thinking, Fast and Slow.* Farrar, Straus and Giroux, 2011.

Kaplan, Michael and Ellen Kaplan. *Chances Are: Adventures in Probability.* Penguin Books, 2006.

Kelly, Henry et al. "Flying Blind: The Rise, Fall and Possible Resurrection of Science Policy Advice in the United States." Federation of American Scientists, Occasional Paper no. 2, December 2004.

Kennedy, Roger G. *Wildfire and Americans.* Hill & Wang, 2006.

Kleinman, Daniel Lee et al., eds. *Controversies in Science and Technology.* University of Wisconsin Press, 2005.

Koomey, Jonathan G. *Turning Numbers into Knowledge.* Analytics Press, 2001.

Krimsky, Sheldon. *Science in the Private Interest.* Rowman & Littlefield, 2003.

Kuhn, Thomas S. *The Structure of Scientific Revolutions.* University of Chicago Press, [1962] 2012.

Lakoff, George. *The Political Mind: A Cognitive Scientist's Guide to Your Brain and Politics.* Viking, 2008.

Lebo, Lauri. *The Devil in Dover.* New Press, 2008.

Luntz, Frank. *Words That Work.* Hyperion, 2007.

Mann, Michael E. *The Hockey Stick and the Climate War.* Columbia University Press, 2012.

McGarity, Thomas D. and Wendy E. Wagner. *Bending Science: How Special Interests Corrupt Public Health Research.* Harvard University Press, 2008.

McGrayne, Sharon Bertsch. *The Theory that Would Not Die.* Yale University Press, 2011.

Merton, Robert. *On Social Structure and Science.* University of Chicago Press, 1996.

————. *The Sociology of Science.* University of Chicago Press, [1938] 2006.

Michaels, David. *Doubt Is Their Product.* Oxford University Press, 2008.

Miller, Kenneth. *Finding Darwin's God.* Harper Perennial, 1999.

Mooney, Chris. *The Republican War on Science.* Basic Books, 2005.

National Institute for Trial Advocacy. *Federal Rules of Evidence.* National Institute for Trial Advocacy, 2003.

National Research Council. *Reference Manual on Scientific Evidence.* 3rd ed. National Academies Press, 2011.

Nestle, Marion. *Food Politics.* University of California Press, 2002.

Offit, Paul A., M.D. *Autism's False Prophets.* Columbia University Press, 2008.

Oreskes, Naomi and Erik M. Conway. *Merchants of Doubt.* Bloomsbury Press, 2010.

Park, Robert. *Voodoo Science.* Oxford University Press, 2000.

Paulos, John Allen. *Innumeracy: Mathematical Illiteracy and Its Consequences.* Hill and Wang, 2001.

————. *A Mathematician Reads the Newspaper.* Basic Books, 1995.

Pigliucci, Massimo. *Nonsense on Stilts.* University of Chicago Press, 2010.

Pilkey, Orrin H. and Linda Pilkey-Jarvis. *Useless Arithmetic.* Columbia University Press, 2007.

Pollan, Michael. *Food Rules: An Eater's Manual.* Penguin Books, 2009.

————. *The Omnivore's Dilemma: A Natural History of Four Meals.* Penguin Books, 2006.

Reynolds, Handel, M.D. *The Big Squeeze.* Cornell University Press, 2012.

Ricks, Thomas E. *Fiasco.* Penguin Books, 2007.

Ropeik, David. *How Risky Is It Really?* McGraw Hill, 2010.

Seethaler, Sherry. *Lies, Damned Lies and Science.* Pearson Education, 2009.

Seife, Charles. *Proofiness.* Viking, 2010.

Shulman, Seth. *Undermining Science: Suppression and Distortion in the Bush Administration.* University of California Press, 2006.

Slovik, Paul. *The Perception of Risk.* Earthscan, 2001.

Snow, C. P. *The Two Cultures.* Cambridge University Press, 1998.

Steinberg, Ted. *Acts of God: The Unnatural History of Natural Disasters in America.* Oxford University Press, 2000.

Stevens, William K. *The Change in the Weather.* Random House, 1999.

Sunstein, Cass. *Worst-Case Scenarios.* Harvard University Press, 2007.

Tickner, Joel A., ed. *Precaution.* Island Press, 2003.

Wagner, Wendy and Rena Steinzor. *Rescuing Science from Politics.* Cambridge University Press, 2006.

Weinberger, Sharon. *Imaginary Weapons.* Nation Books, 2006.

Woloshin, Steven, M.D., et al. *Know Your Chances.* University of California Press, 2008.

ACKNOWLEDGMENTS

Many people helped me produce this book.

I owe an immense debt to the Shorenstein Center on Media, Politics, and Public Policy at the Kennedy School of Government at Harvard, whose fellowship enabled me to begin work on this project more than a decade ago. Alex Jones, my onetime *New York Times* colleague and the center's longtime director, made my fellowship there possible. Thank you, Alex. And thanks also to Edie Holway, who directed the fellowship program. The fellowship was a rare gift, and I am grateful for it.

James J. McCarthy, the Alexander Agassiz Professor of Biological Oceanography at Harvard, invited me to teach a seminar on the book's theme while I was in Cambridge, and Daniel P. Schrag, the Sturgis Hooper Professor of Geology, renewed the teaching invitation through the Harvard University Center for the Environment, which he directs. Discussions there were always fascinating. At HUCE, James Clem, Lisa Matthews, and Jean Gaultier are among the many other people whose kindness and helpfulness contributed so much to my experiences there. Working with them has been immensely rewarding and great fun.

I am also grateful to the people at Brown who encouraged me to teach there, including Clyde Briant, the Otis E. Randall University Professor of Engineering and then the university's vice president for research; Mark Schlissel, then Brown's provost; and Professor of Engineering and Senior Associate Dean of the Faculty Janet Blume, who as interim director of the Brown Center for Environmental Studies was an enormous help to me. (She also recruited me to sing in her garage band, which has been a blast!)

At the Institute at Brown for Environment & Society, I have been grateful for the assistance and guidance of Amanda Lynch, its director and Professor of Earth, Environmental, and Planetary Sciences, and Dov Sax, its deputy director and Professor of Ecology and Evolutionary Biology. My other colleagues in the Center for Environmental Studies have been unfailingly helpful (and patient). Thanks especially to Jeanne Lowenstein and Kurt Teichert.

The Brown and Harvard students who have taken my courses have provided lively feedback, raised interesting questions, and otherwise contributed to this project. Along the way, some of them have also instructed me on the ways of cooking, Powerpoint, Dropbox, and other useful technologies. Thanks to all of you!

It is impossible to overestimate the debt of gratitude I owe my professional colleagues in journalism at the *Providence Journal* and especially at the *New York Times*. Working at both newspapers has been an education, in the best sense of the word. In particular, I am overwhelmed by affection and esteem for my colleagues in the *Times*'s Science Department, and I will always be grateful to Joseph Lelyveld, who as the newspaper's executive editor gave me the chance to lead it.

Michael Fisher of Harvard University Press helped me shape this project and exhibited superhuman patience as I missed one deadline after another. (I had good excuses, but still.) He always had faith in this work and for that I am very grateful. I thank Andrew Kinney, who inherited the project from Michael, and Katrina Vassallo and Kate Brick. I am grateful for their attention (and their meticulous editing). Thanks also to the designer, Peter Holm / Sterling Productions, and Margaux Leonard, my publicist at Harvard University Press.

As always, my agent Jim Levine has offered encouragement and useful advice. Thanks Jim!

Finally—and most importantly—I must acknowledge the scientists, engineers, and other researchers who over the years have helped me and my colleagues report on the research enterprise from our orbits on its periphery. This work has always been worthwhile and almost always it has been the best possible fun. Thank you!

INDEX